Bio Lubrication for Sustainable Tribology

With the growing demand for sustainable solutions in engineering, biolubricants are at the forefront of modern tribology. *Bio Lubrication for Sustainable Tribology* explores the latest advancements in bio-based lubricants, green additives, and stabilization techniques, shedding light on their role in enhancing surface interactions across various applications. It provides crucial insights into the future of eco-friendly lubrication and its impacts in terms of efficiency, longevity, and sustainability including environmental responsibility.

This book:

- Presents the latest research developments and recent trends such as bio-additives, nano greases, nanocomposite tribology, and the concept of bio-tribology within the context of sustainable development.
- Highlights sustainability, tribology of biological systems, and chemical aspects of bio-coatings.
- Focuses on discussing the role of biomaterials in promoting sustainability and covers topics such as improving wear characteristics.
- Explains the tribology of bio-lubricants, bio-additives, lightweight bio-fillers, and reinforcements materials.
- Illustrates topics such as orthopedic implants, and biocompatible nanomaterials related to the role of nanomaterials in bio-tribology.

It is primarily written for senior undergraduate, graduate students, and academic researchers in disciplines including mechanical engineering, manufacturing engineering, industrial and production engineering, automotive engineering, and aerospace engineering.

Bio Lubrication for Sustainable Tribology

Edited by
Zulfiqar Khan,
M. S. Charoo,
Pranav Dev Srivyas, and
Tanmoy Medhi

CRC Press
Taylor & Francis Group
Boca Raton London New York

CRC Press is an imprint of the
Taylor & Francis Group, an **informa** business

Designed cover image: OpenAI's DALL·E

First edition published 2026
by CRC Press
2385 NW Executive Center Drive, Suite 320, Boca Raton FL 33431

and by CRC Press
4 Park Square, Milton Park, Abingdon, Oxon, OX14 4RN

CRC Press is an imprint of Taylor & Francis Group, LLC

ISBN: 978-1-032-65991-6 (hbk)
ISBN: 978-1-041-20642-2 (pbk)
ISBN: 978-1-003-71736-2 (ebk)

DOI: 10.1201/9781003717362

Typeset in Sabon
by SPi Technologies India Pvt Ltd (Straive)

Contents

About the Editors

Zulfiqar Khan has been leading the NanoCorr, Energy and Modeling (NCEM) research group. He has developed and established a multidisciplinary research portfolio in Nanoengineering and Energy Systems (NES®) in collaboration with major industry and HEI partners. Professor Khan and his team's technological innovations are evidenced by several GB and international patents in corrosion condition monitoring, low- to zero-emission energy technologies, and nanofluids. His research interests include nanocomposites, wear-corrosion synergy, surface engineering, sustainable design, and low-carbon technologies. He is the recipient of the Distinguished Researcher in Tribology Award, Excellence in Education in Sustainable Development Award, Albert Nelson Lifetime Achievement Award, Doctoral College Outstanding Contribution Award, and Vice Chancellor's staff awards in enriching society and advancing knowledge. He is a Fellow of the Institution of Mechanical Engineers (IMechE), Chartered Engineer (CEng) and Senior Fellow of the Higher Education Academy (HEA) UK. He is an editorial board member of several high-ranking journals.

M. S. Charoo (PhD) is an Associate Professor at the Department of Mechanical Engineering, National Institute of Technology Srinagar (J&K), India. Professor Charoo has over 25 years of teaching experience both at postgraduate and undergraduate levels in the areas of Lubrication, Design, and Tribology. A lifetime member of the Tribology Society of India, he has co-authored over 75 research articles in international journals and conference proceedings of high repute. He has supervised four PhD scholars along with numerous MTech dissertations. He serves on the reviewer panel of many international journals.

Pranav Dev Srivyas holds a Ph.D. in Mechanical Engineering from the National Institute of Technology (NIT) Srinagar, with a specialization in tribology, focusing on self-lubricating composites and nano-lubrication. He further advanced his research career through a Post-Doctoral Fellowship at the Indian Institute of Technology (IIT) Guwahati, where he explored smart fluids and metamaterials. With a unique blend of academic and industrial experience, Dr. Srivyas is currently associated with Ind-Sphinx Precision Ltd. (Axis Tools), a leading manufacturer in the cutting tool industry. His multidisciplinary research interests span surface engineering, coating technologies, smart fluids, composite materials, laser texturing, and advanced cutting tools. He has contributed extensively to the field through numerous publications in reputed international SCI journals, establishing himself as a dedicated researcher and innovator in mechanical and materials engineering.

Tanmoy Medhi is working as an Assistant Professor in the Department of Mechanical Engineering, School of Engineering, Tezpur University. He holds a Ph.D. in Mechanical Engineering from NIT Agartala, where his research focused on friction stir welding of dissimilar Al-Cu joints. With a strong foundation in welding science, solid-state additive manufacturing, and hybrid manufacturing processes, Dr. Medhi has contributed extensively to both academic research and industrial applications. He also carried out independent research in the areas of bio-lubrication and sustainable manufacturing while working as a Post-Doctoral Fellow in the Department of Mechanical Engineering at the Indian Institute of Technology Guwahati. He has led and collaborated on funded projects with agencies such as SERB, ISRO, DST, and Shree Cement Limited. He holds a patent for an innovative induction heating-assisted friction stir welding technique and serves as an Academic Editor for journals such as *Scientific Reports* (Nature Portfolio) and the *Journal of Engineering* (Hindawi). His academic excellence is further reflected by awards such as the M.Tech Gold Medal and the prestigious National Post-Doctoral Fellowship from SERB. Actively involved in academic development, he has coordinated workshops and short-term training programs for defense and industry stakeholders. Dr. Medhi is also a professional member of the Indian Welding Society and the Institution of Engineers India.

Contributors

Ashish Kumar Sinha
Department of Mechanical Engineering Department, National Institute of Technology Srinagar
Hazratbal, Srinagar (J&K), India

Bharat Kumar
Center for Automotive Research and Tribology, Indian Institute of Technology Delhi
New Delhi, India

Chandra Shekhar
Tribology Laboratory, National Institute of Technology Srinagar
Hazratbal, Srinagar (J&K), India

Chattar Pal
Ind-Sphinx Precision Ltd.
Parwanoo, Himachal Pradesh, India

Chennupati Lakshmi Srinivas
Bapatla Engineering College
Bapatla, Guntur, Andhra Pradesh, India

D. Amrishraj
Department of Mechanical Engineering, V.S.B. Engineering College
Karur, Tamil Nadu, India

D. Dsilva Winfred Rufuss
School of Mechanical Engineering, Vellore Institute of Technology
Vellore, Tamil Nadu, India

Dr. Jagtar Singh
Spinks World Printing Technologies
Baddi, Himachal Pradesh, India

Durga Prasad Kelli
Aditya Institute of Technology and Management (A)
Tekkali, Andhra Pradesh, India

Grandhi Prasanth
Department of Mechanical Engineering, NIT Andhra Pradesh
Tadepalligudem, Andhra Pradesh, India

Gunji Srinivasa Rao
RVR&JC College of Engineering
Chowdavaram, Guntur, Andhra Pradesh, India

Hymavathi Madivada
Department of Mechanical Engineering, NIT Andhra Pradesh
Tadepalligudem, Andhra Pradesh, India

Himanshu Chandra
Indian Institute of Technology Kharagpur
Kharagpur, West Bengal, India

Himanshu Shekhar Gupta
Department of Mechanical Engineering Department, National Institute of Technology Srinagar
Hazratbal, Srinagar (J&K), India

J. Anita Jessie
Department of Civil Engineering, Vel Tech Rangarajan Dr. Sagunthala R&D Institute of Science and Technology
Avadi, Chennai, Tamil Nadu, India

Jayakrishna Kandasamy
Department of Manufacturing Engineering, School of Mechanical Engineering, Vellore Institute of Technology
Vellore, Tamil Nadu, India

Jissy Jacob
International and Inter University Centre for Nanoscience and Nanotechnology, Mahatma Gandhi University
Kottayam, Kerala, India

R. Jose Immanuel
IITBH
Bhilai, Chhattisgarh, India

Kelli Durgaprasad
Department of Mechanical Engineering, Aditya Institute of Technology and Management (A)
Tekkali, Andhra Pradesh, India

M. Hanief
National Institute of Technology Srinagar
Hazratbal, Srinagar (J&K), India

M. Natesh
Department of Mechanical Engineering, V.S.B. Engineering College
Karur, Tamil Nadu, India

M. F. Wani
Department of Mechanical Engineering Department, National Institute of Technology Srinagar
Hazratbal, Srinagar (J&K), India

M. S. Charoo
Department of Mechanical Engineering Department, National Institute of Technology Srinagar
Hazratbal, Srinagar (J&K), India

Mukund Dutt Sharma
Department of Mechanical Engineering Department, National Institute of Technology Srinagar
Hazratbal, Srinagar (J&K), India

P. Jeyapandiarajan
School of Mechanical Engineering, Vellore Institute of Technology
Vellore, Tamil Nadu, India

Pankaj Biswas
Department of Mechanical Engineering, Indian Institute of Technology
Guwahati, Assam, India

Pranav Dev Srivyas
Ind-Sphinx Precision Ltd., Parwanoo, Himachal Pradesh 173220, India.

Priyam Kashyap
NERIST & IITBH
Bhilai, Chhattisgarh, India

Prof. Umida Ziyamukhamedova
Faculty of Railway Engineering, Tashkent State Transport University
Tashkent, Uzbekistan

R. Prayer Riju
School of Mechanical Engineering, Vellore Institute of Technology
Vellore, Tamil Nadu, India

Rakesh Sehgal
Department of Mechanical Engineering, National Institute of Technology Hamirpur
Himachal Pradesh, India

Roopsandeep Bammidi
Department of Mechanical Engineering, Aditya Institute of Technology and Management (A)
Tekkali, Andhra Pradesh, India

S. Aadil Hussain
Department of Mechanical Engineering, National Institute of Technology Srinagar
Hazratbal, Srinagar (J&K), India

S. Arulvel
School of Mechanical Engineering, Vellore Institute of Technology
Vellore, Tamil Nadu, India

Sandeep Soni
Tribology Laboratory, Department of Mechanical Engineering, S.V. National Institute of Technology
Surat, Gujarat, India

Sanjay Kumar
Department of Mechanical Engineering, National Institute of Technology Srinagar
Hazratbal, Srinagar (J&K), India

Santosh Kumar Tamang
NERIST
Nirjuli, Arunachal Pradesh, India

Shariq Ahmad Bhat
Department of Mechanical Engineering, National Institute of Technology Srinagar
Hazratbal, Srinagar (J&K), India

Sheikh Haris Mukhtar
Tribology Laboratory, National Institute of Technology Srinagar
Srinagar, J&K, India

Sivasankara Raju Rallabandi
Aditya Institute of Technology and Management (A)
Tekkali, Andhra Pradesh, India

Soundhar Arumugam
Department of Mechanical Engineering, Vel Tech Rangarajan Dr. Sagunthala R&D Institute of Science and Technology
Avadi, Chennai, Tamil Nadu, India

Sreeramulu Dowluru
Department of Mechanical Engineering, Aditya Institute of Technology and Management (A)
Tekkali, Andhra Pradesh, India

Suraj Tiwari
CIT Kokrajhar
Kokrajhar, Assam, India

Tanmoy Medhi
Department of Mechanical Engineering, Indian Institute of Technology
Guwahati, Assam, India

Venkatesh Muddada
Department of Mechanical Engineering, Aditya Institute of Technology and Management (A)
Tekkali, Andhra Pradesh, India

Wani Khalid Shafi
Islamic University of Science and Technology
Awantipora, Pulwama, Jammu-Kashmir, India

Chapter 1

Introduction to bio-lubricants

Tanmoy Medhi, Pranav Dev Srivyas, Soundhar Arumugam, and Pankaj Biswas

1.1 INTRODUCTION

Lubricants are employed on moving surfaces in numerous applications all over the world. Their primary function is to lessen heat and wear between surfaces of relative motion, but with the understanding that wear cannot be totally eliminated, extending the life of the surfaces has become an important secondary objective. Along with this, some other objectives of lubrication, such as noise reduction, power saving, and surface finish preservation, have evolved as a result of the needs and demands of the modern industrial era. These factors are particularly important in specialized applications with a focus on the likes of automotive and cutting-tool industries. An understanding of the detrimental effects that can be caused by a lack of suitable lubrication (such as machine seizure and bearing failure) has highlighted the need for an effective maintenance program. The importance of lubrication to effective maintenance and the reduction of overall operating costs at a time when worldwide economies are under pressure should not be underestimated.

From the early days of human civilization, it has been realized that reducing friction between moving surfaces to an absolute minimum is beneficial. The accepted method of achieving this has become known as lubrication, but rapid advances in technology and the ever-increasing demand for energy conservation have meant that the science and practice of lubrication have become a vital and complex aspect of modern industrial culture. Originally, lubrication had the simple goal of minimizing the friction between moving parts; however, with the expansion of different types of machines, the higher expectations for lower energy consumption have expanded the purpose and scope of lubrication, and it has also raised the understanding of the economic relevance of proper lubrication practices. Lubrication is considered an essential step in the effective elimination of friction in today's modern machinery. This, in turn, helps to increase its wear lifetime and, in many cases, the rate of energy conserved during operations. It has been determined that using a good lubricant can reduce the energy used by two rubbing or interacting surfaces under harsh conditions. This is because both a

DOI: 10.1201/9781003717362-1

significant improvement in the machine's performance and a reduction in the amount of heat produced by the two components in contact are possible. In situations that involve parts with extremely tight tolerances, lubrication is often the only way in which the parts can be operated and, in some cases, the only means of ensuring their assembly. A well-lubricated machine or mechanism is more likely to be reliable. It has been said that a good design can become a poor one in the absence of correct lubrication. As a matter of interest, it has been estimated that between 10 and 20% of all mechanical failures are the direct result of poor lubrication or contamination of lubricated parts. In severe cases, certain materials in relative motion simply weld together if there is no lubrication or if it is lost. This is common in some metals due to their naturally high adhesion. The removal of such parts can be quite difficult, particularly if they have failed in an inaccessible location for tools.

The lubricants are broadly categorized into three types: (a) Petroleum-based lubricants (mineral oils); (b) natural oils (vegetable oil or animal fat-based); and (c) synthetic oils. Mineral oils are basically petroleum-based lubricants, also termed as conventional lubricants, which are sourced from crude oil, are not renewable, and represent substantial threats to the environment. The modern world is strongly reliant on machinery and equipment, ranging from industrial machines to automobile engines, all of which require good lubrication to function smoothly and efficiently. Petroleum-based lubricants have traditionally been used to meet these lubrication needs due to their superior performance, cost-effectiveness, and availability. However, the use of conventional lubricants has come under increased criticism due to their environmental and health implications. Their extraction, manufacture, and disposal all contribute to pollution, including soil and water contamination, emissions of greenhouse gases, and environmental deterioration. As these concerns become more severe, the desire for sustainable solutions has led to the resurgence of lubricants that are derived from bio-based feedstock in industrial applications in the last three decades [1]. As the theme of the book is aligned toward bio-lubrication, this chapter will present a basic introduction about the same.

1.2 DEFINITION AND ORIGIN OF BIO-LUBRICANTS

Bio-lubricants are a type of lubricant that are generated from renewable biological sources such as vegetable oils, animal fats, and other environmentally compatible hydrocarbons. These constituents serve as a base for bio-lubricants, which are frequently supplemented with additives to improve performance. Unlike petroleum-based lubricants, bio-lubricants are biodegradable and nontoxic, making them a more environmentally friendly option [2]. The effect of using traditional lubricants such as their impact on the environment, health risks, and the nonrenewable nature of fossil fuels, has greatly influenced the development and adoption of bio-lubricants [3, 4]. The bio-lubricants are mainly classified as natural and synthetic oils. The natural oil comprises vegetable oil (soybean oil, rapeseed oil, sunflower oil, palm oil, castor oil, etc.) and animal fat (tallow, lard, etc.) as base oil. The synthetic oil also uses natural oil as a base material but is chemically modified by the esterification process to enable the grafting of microorganisms, alcohols, polyglycols, polyalcohols, perfluoroalkyl ethers, etc., into natural oil in order to create more sophisticated bio-lubricants. In addition to the base oils, bio-lubricants often incorporate various additives (butylated hydroxytoluene (BHT), organic phosphates, olefins, polymethacrylates, etc.,) to enhance their performance and functionalities. These additives can improve properties such as wear resistance, viscosity, and oxidative stability, making the bio-lubricants suitable for demanding applications. The detailed discussion of the types of methodology to formulate and modify will be presented in the successive chapters.

The primary advantage of bio-lubricants lies in their environmental benefits. These are typically more biodegradable than their petroleum-based counterparts, reducing the risk of long-term contamination of soil and water bodies. Bio-lubricants are probably less toxic to wildlife, aquatic organisms, and humans as compared to conventional oil. In addition, the synthesis and utilization of the bio-lubricants are normally associated with the restricted, or even lower, emissions of greenhouse gases, thus leading to the decrease of the carbon footprint – something even more significant in the case where the production of the bio-lubricants is based on the vegetable raw materials grown in a sustainable manner. Other performance characteristics of bio-lubricants include improved lubricity, and as a consequence, lower friction and wear of the equipment, which means longer service life and increased output. Bio-lubricants may also possess a high viscosity index, implying that they are more stable at high and low temperatures compared to when they are synthesized as individual products. Furthermore, bio-lubricants usually exhibit lower volatility than petroleum-based products, and thus, there is less loss due to evaporation when the product is in use and the risk of fire and containment also poses a threat to the health of people in the working environment [5]. These advantages, however, are accompanied by certain areas that would require some distinct procedures to be accomplished to ensure broader adoption. Bio-lubricants remain a disadvantage since they are expensive to manufacture at first and distribute across the market. This is because the raw materials involved and the processes that are needed to achieve some potential properties make the product more expensive than petroleum-based lubricants. Therefore, increased manufacturing costs push the price tag higher, meeting the market, which acts as a massive barrier to implementation, mainly for industries that cannot afford such prices. Bio-lubricants may also show some reduction in overall performance as compared to conventional lubricants. For example, they might display lower oxidative stability, which in turn can make their products have a shorter shelf life and poor performance at high temperatures [6]. However, some of the bio-lubricants are relatively more susceptible to hydrolysis (break down on interaction with water); this may be an issue where they are likely to come into contact with water. Furthermore, bio-lubricants are known to have low cold flow properties, whereby they reach high viscosity at cold temperatures unless the problem is corrected by adding certain compounds [7]. In the supply chain and availability of bio-lubricants, there are also some challenges involved. In the case of the raw materials for bio-lubricants, crop yields, time of the year, competitive uses of the crop, and variations in yields are very significant factors that may affect availability. Another disadvantage is that they have compatibility problems with the machinery, seals, and hoses used in existing equipment and may degrade or cause failure of these components [8]. Finally, the general ignorance of the technical feasibility of bio-lubricants is always a serious drawback preventing further market expansion.

1.3 HISTORICAL CONTEXT OF BIO-LUBRICANTS

The use of lubricants has a long and storied history, going back to ancient civilizations. Lubricants can be traced back to the earliest-known liquid that the cultures of that time used, including animal fats and vegetable oils, among other products, which included the use of olive oil in the axle of chariots of Egypt around 1400 BC [9]. Likewise, contemporary prehistoric cultures employed olive oil and other plant-based oils for such uses [10].

The Industrial Revolution came toward the end of the 18th century and was accompanied by several changes in mechanical and manufacturing technology, and it called for higher quality and improved lubricants. Since crude oil was discovered and exploited, products made from petroleum have replaced those made from natural oils and fats. The use of crude oils in the mid-19th century to produce mineral oils was also uncovered, which served as a much better fluid for lubrication than the natural oils [8]. The application of vegetable oils and animal fats as a lubricant continued until early in this 20th century [6]. But with the progress of the 20th century, the use of mineral oil gained pace and became the standard in industrial and automotive applications due to its availability, cost-effectiveness, and enhanced performance characteristics such as higher stability, better viscosity control, and longer service life [11]. This gradually overshadowed the utilization of natural oils, which limited its research and development and restricted its use to niche areas like food processing industries where biodegradability and nontoxic properties were required [12]. The use of extensive mineral oils led to the contamination of the environment, soil, and water due to their nonbiodegradable nature. In addition, the frequent extraction and refining of crude oil contributed to other major ecological impacts. It was in 1970 that the situation of the oil crisis emphasized the dangers of relying too much on petroleum resources, encouraging research into alternative, renewable lubricants [13]. The need for more eco-friendly and sustainable solutions reignited interest in bio-lubricants. These prompted increased research and development efforts to improve the performance and cost-effectiveness of bio-lubricants, which extensively took place in 1980 and 1990 [12]. This period saw advances in the extraction, refining, and formulation of bio-based lubricants, as well as the investigation of additional feedstock sources such as vegetable oils and animal fats [14, 15]. The current phase, which has accelerated in the 21st century, is distinguished by widespread recognition of the environmental benefits of bio-lubricants, as well as increasingly severe legislation and customer demand for eco-friendly products [13]. This has resulted in additional scientific improvements, the formation of industry standards, and the spread of bio-lubricant applications across a variety of industries, including automotive, industrial, and maritime. As the world transitions to a more sustainable future, the continued development and adoption of bio-lubricants are expected to play a critical role in reducing the environmental impact of industrial and transportation activities, while providing viable alternatives to traditional petroleum-based lubrication [6, 13, 16].

1.4 PROPERTIES OF BIO-LUBRICANTS

Bio-lubricants reduce friction and wear, disperse deposits, prevent rust and corrosion, dissipate heat, and seal contact joints. To perform these vital functions effectively, the base oil of a bio-lubricant must have appropriate viscosity and viscosity index, low volatility to minimize deposit formation, good performance at low temperatures, and strong hydrolytic and thermo-oxidative stability, as well as being biodegradable. The general properties are essential to understand, which are described as follows:

1. Viscosity: Viscosity is one among the various parameters that have to be assessed for the transition from the normal to the bio-lubricants because the viscosity determines their functionality and compatibility within different applications. Researchers also observe that bio-lubricants, usually produced from renewable resources including vegetable oil and animal fats, have comparatively high viscosity with petroleum-based lubricants [17]. This characteristic may be attributed to the molecular characteristics, especially the hydrocarbon chain length of bio-lubricant constituents, and the presence of functional groups like hydroxyl. The increase in viscosity increases the mobility and decreases friction, but the focus has to be achieved with the aptitude of the targeted application. Some studies indicated that the temperature affects the viscosity of bio-lubricants as well [18]. Bio-lubricants with a higher viscosity index are normally preferred based

on the consideration that their viscosity changes less with variation in temperature compared to those with a lower viscosity index, and hence, they suit a range of operating temperatures.

2. Thermo-oxidative stability: This property defines the response of bio-lubricants to heat and exposure to oxygen when in use. The level of thermo-oxidative stability of bio-lubricants depends on the fatty acid profile of the base stock. Bio-lubricants developed from vegetable oils are found to undergo thermal and oxidative degradation at a faster rate than their mineral oil counterparts because the base oils used contain unsaturated fatty acids that readily oxidize at high temperatures. However, to improve the thermo-oxidative stability of bio-lubricants, chemical modification of vegetable oil is also possible; for instance, using epoxidation to saturate the carbon-carbon double bonds (–C=C–). From observation, it has been realized that this method enhances both the thermo-oxidative stability and viscosity index of the bio-lubricant.
3. Hydrolytic stability: The hydrolytic stability is another important factor to consider for bio-lubricant stability that accounts for their resistance to chemical attacks involving water as a reactant or product. Bio-lubricants are typically composed of esters, which can undergo hydrolysis through the action of carboxylic acid and alcohol groups. Studies have indicated that enhancing saturation contributes to improved hydrolytic stability, with saturated dicarboxylic esters demonstrating high stability due to steric effects. The use of short-chain alcohols has been found to enrich the hydrolytic stability of bio-lubricants, and employing branched alcohols can further enhance this resistance.
4. Pour point: Pour point of a bio-lubricant is the parameter that determines its ability to flow at low temperatures. The ester group in bio-lubricants provides low-temperature fluidity and decreases volatility at elevated temperatures, making them favorable for uses that prioritize chemical stability [19]. However, the pour point of bio-lubricants can be a challenge, as the typical polyunsaturated fatty acids found in most vegetable oils tend to solidify at low temperatures. The ASTM D5949 standard procedure is frequently utilized in the determination of lubricants' pour point. This method includes lowering the temperature of the lubricant sample while introducing a sudden release of nitrogen gas, and employing an optical instrument to observe the shifting of the sample's surface as it undergoes a reduction in its ability to flow.
5. Eco-toxicity: Eco-toxicity is another index of the environmental toxicity of a lubricant formulation since it identifies whether a lubricant is capable of making life forms permanently dysfunctional. Lubricant water toxicity is also relevant to watershed environments because the toxicity of water-soluble lubricant components, such as the organics

and hydrocarbons, is influential to lubricant water toxicity. The LD_{50} metric is often used to measure toxicity, indicating the dose needed to cause fatality in 50% of a population. As per scholarly sources, a bio-lubricant is deemed nontoxic to the environment if its LD_{50} surpasses 1000 ppm [1].

6. Bio-degradability: It is the capability of a bio-lubricant to break down into the form of renewable raw materials through aerobic or anaerobic processes by organisms or parent enzymes. The process of bio-degrading a lubricant involves multiple stages. Initially, the original bio-lubricant undergoes transformation to produce a new compound, which may or may not be completely biodegradable. This primary breakdown is assessed by analyzing the C–H bond using infrared spectroscopy. The subsequent stage includes breaking down the organic compound into CO_2 and H_2O through biodegradation within 28 days [20]. However, assessing biodegradability can be difficult due to potential variation in results from different testing methods.

1.5 APPLICATIONS OF BIO-LUBRICANTS

The application of lubricants can be categorized into two distinct systems: closed and open. In closed systems, lubricants are designed to be contained and insulated, preventing uncontrolled release into the environment. Conversely, open systems involve the continuous release of lubricants into the surrounding environment. However, even in closed systems, mishandling or human error can lead to the inadvertent disposal of lubricants, resulting in undesired environmental contamination [9]. Given these considerations, the development of alternative lubricants to traditional mineral oils must prioritize high levels of biodegradability to mitigate environmental impact in the event of unintended release. Hence, the properties of the bio-lubricants must be known before intending to use them for a specific purpose (Table 1.1).

Hydraulic oil: Hydraulic fluids are used in transferring energy and power within hydraulic systems, as well as play a major role in lubricating these systems. One important characteristic of hydraulic fluids is their compressibility, as fluids with lower compressibility can enhance pressure transmission speed, resulting in quicker response times and more efficient energy utilization. For instance, vegetable oils like soybean oil have been observed to exhibit low isothermal compressibility and suitable viscosity for use in hydraulic systems. Moreover, they also demonstrate the ability to remove air bubbles from the fluid at a rate seven times faster than the required standards. In addition, various other vegetable oils, such as rapeseed oil, palm oil, moringa oil, passion fruit oil, and rubber seed oil, have displayed properties that are comparable

Table 1.1 Some applications of bio-lubricants with their properties [6]

Application	*Properties*	*Advantages*
Engine oil	Minimum emission of volatile organic compounds, good lubricating properties	Reduces emission from the engine increases the performance of the engine
Hydraulic oil	Low values of compressibility, high rates of time to discharge air	Improved pressure transmission, fewer vibrations and noises
Compressor oil	High thermal resistance	Endures temperature and pressure above the standards of standard atmospheric pressure
Metalworking oil	Low flakiness, good antirust performance, good emulsifying property, good lubricating property	Reduced generation of mist, increased life of the tool, heat-stable emulsions
Transmission oil	Ease of sliding, greater weld force	Impartible additives can then be added
Chainsaw oil	Low volatility	Reduced hazardous mist generation
Insulating oil	Increased amount of water solubility and high dielectric constant	Reduces the influence of moisture on insulation strength, enhanced insulation characteristics

to those of traditional mineral oils. The combination of vegetable oils with alcohols and acids has been documented to enhance the performance of hydraulic fluids, leading to improved biodegradability and a reduced pour point. Leading companies such as *Shell, Mobil*, and *Chevron Texaco* are currently promoting hydraulic fluids and systems that utilize environmentally sustainable vegetable oils meeting different pump test standards.

Metalworking oil: Metalworking fluids are important lubricants that are utilized in machining to cool and lubricate chips and edges, thus averting system heating. Nonetheless, the disposal of these fluids raises environmental and health issues, especially due to the risk of inhalation. Vegetable-derived oils have surfaced as a hopeful option for addressing these challenges. Soybean oil has a reduced environmental footprint in comparison to traditional metalworking fluids, but it may release volatile compounds due to its thick consistency and low evaporation rate. According to Syahir et al., the use of bio-based vegetable oils in manufacturing metalworking fluids led to improvements in surface smoothness, tool wear, and cutting temperatures. Since these fluids are often emulsified with water, factors such as rust prevention and emulsifiability need to be taken into account. Soybean oil esterification and adjustments to karanja and coconut oils have been documented as methods to enhance the thermal stability as well as resistance to oxidation of the metalworking fluids. Moreover, the addition of bio-based palm oil has been demonstrated to increase metal removal rates in steel turning processes.

Greases: Another application area is the use of bio-lubricants as a grease for machinery and equipment, which can provide several advantages over conventional greases. The bio greases need to be mixed with a thickener, that is basically a lithium, compound that exhibits enhanced lubricity, high viscosity, good anti-wear characteristics, and a high viscosity index – all of which contribute to improved equipment performance and service life. In addition, greases derived from biological sources have minimal evaporation rates and release fewer metal residues into the air, rendering them a more eco-friendly alternative.

High-performance oil: These are specialized lubricants called performance lubricants that are developed to fulfill the challenging needs of specific high-performance uses. They are designed to offer exceptional lubrication qualities under severe conditions, including elevated temperatures, extreme pressures, and harsh surroundings. Performance lubricants can be synthetic or produced through advanced refining methods and aim to improve the effectiveness, dependability, and lifespan of machinery and components. Some areas where performance lubricants are used are transmission fluids, racing vehicles, aircraft engines and components, and ship engines and systems.

Penetrating oil: Penetrating oils or penetrants are specifically formulated lubricants intended to permeate confined areas, release rusted or immovable components, and offer lubrication. These lubricants contain rust inhibitors, corrosion inhibitors, anti-wear compounds, and, in certain cases, mild solvents to improve penetration and efficacy.

1.6 CONCLUSION

It can be concluded that vegetable oil lubricants are well-liked for their excellent lubricating properties, low toxicity, environmental friendliness, and ability to biodegrade. However, their use in cooking can increase costs and impact availability on a large scale. Nonconsumable oils also offer an intriguing prospect in terms of waste handling and obtaining importance from byproducts. In this regard, the utilization of bio-oils produced from lignocellulosic agro-residues, microalgae, and plastic-rich fractions like refuse-derived fuels using thermochemical methods such as pyrolysis and HTL requires further attention. It is evident that bio-lubricants offer superior lubrication properties compared to commonly used synthetic and mineral oils. They also possess the necessary attributes for diverse engineering applications. Furthermore, their nontoxic and clean nature makes them suitable for environmentally sensitive uses in marine and forestry ecosystems, subject to stringent laws and regulations. Consequently, there is anticipated growth in the demand for bio-lubricants across various applications, including hydraulic oils, gear oils, engine oils, cutting fluids, turbo-machinery, and electrical appliances. The shift to bio-lubricants represents a noteworthy advancement in sustainable industrial methods, providing various advantages and effectively addressing significant environmental and health issues. With ongoing technological progress and improved economies of scale, the utilization of bio-lubricants is poised to become more feasible and prevalent in the future.

REFERENCES

1. Reeves CJ, Siddaiah A, Menezes PL. A review on the science and technology of natural and synthetic biolubricants. *Journal of Bio- and Tribo-Corrosion* 2017;3:11.
2. Salimon J, Salih N, Yousif E. Biolubricants: Raw materials, chemical modifications and environmental benefits. *European Journal of Lipid Science and Technology* 2010;112:519–30.
3. Li W, Kong XH, Ruan M, Ma FM, Jiang YF, Liu MZ, et al. Green waxes, adhesives and lubricants. *Philosophical Transactions of the Royal Society A: Mathematical, Physical and Engineering Sciences* 2010;368:4869–90.
4. Lovell M, Higgs CF, Deshmukh P, Mobley A. Increasing formability in sheet metal stamping operations using environmentally friendly lubricants. *Journal of Materials Processing Technology* 2006;177:87–90.
5. Bart JCJ, Gucciardi E, Cavallaro S. Markets for biolubricants. In: *Biolubricants* (eds. E Gucciardi, S Cavallaro, JCJ Bart) Elsevier; 2013. page 712–54.
6. Cecilia JA, Ballesteros Plata D, Alves Saboya RM, Tavares de Luna FM, Cavalcante CL, Rodríguez-Castellón E. An overview of the biolubricant production process: Challenges and future perspectives. *Processes* 2020;8:257.
7. Ho CK, McAuley KB, Peppley BA. Biolubricants through renewable hydrocarbons: A perspective for new opportunities. *Renewable and Sustainable Energy Reviews* 2019;113:109261.
8. Mairiza L, Zuhra, KM, Muhibbul, BI, Yunardi Y. An overview of potential production of bio-lubricant in Indonesia. *IOP Conference Series: Earth and Environmental Science* 2022;951:012061.
9. Mang T, Dresel W, editors. *Lubricants and Lubrication*. Wiley; 2006.
10. Nosonovsky M. Oil as a lubricant in the ancient middle east. *Tribology Online* 2007;2:44–9.
11. Mang T, Gosalia A. Lubricants and their market. In: *Lubricants and Lubrication* (eds. Wilfried Dresel, Theo Mang). Wiley; 2017. page 1–10.
12. Salih N, Salimon J. A review on eco-friendly green biolubricants from renewable and sustainable plant oil sources. *Biointerface Research in Applied Chemistry* 2021;11:13303–27.
13. Salimon J, Salih N, Yousif E. Industrial development and applications of plant oils and their biobased oleochemicals. *Arabian Journal of Chemistry* 2012;5:135–45.

14. Singh Y, Sharma A, Singh N, Singla A, Rastogi PM. Prospects of inedible plant oil-driven bio-lubricants for tribological characteristics – A review. *International Journal of Ambient Energy* 2020;41:1534–47.
15. Karmakar G, Ghosh P, Sharma B. Chemically modifying vegetable oils to prepare green lubricants. *Lubricants* 2017;5:44.
16. Kodali DR. Development, properties and applications of high-performance biolubricants. In: *Advances in Biorefineries* (ed. Keith Waldron). Elsevier; 2014. page 556–95.
17. Chan CH, Tang SW, Mohd NK, Lim WH, Yeong SK, Idris Z. Tribological behavior of biolubricant base stocks and additives. *Renewable and Sustainable Energy Reviews* 2018;93:145–57.
18. Bair S, Michael P. Modelling the pressure and temperature dependence of viscosity and volume for hydraulic fluids. *International Journal of Fluid Power* 2010;11:37–42.
19. Salimon J, Abdullah BM, Yusop RM, Salih N. Synthesis, reactivity and application studies for different biolubricants. *Chemistry Central Journal* 2014;8:16.
20. Battersby NS, Morgan P. A note on the use of the CEC L-33-A-93 test to predict the potential biodegradation of mineral oil based lubricants in soil. *Chemosphere* 1997;35:1773–9.

Chapter 2

Types of bio-lubricants and additives

Jissy Jacob, J. Anita Jessie, M. Natesh, and Soundhar Arumugam

2.1 INTRODUCTION

Lubrication refers to the method or approach of minimizing the wear on one or both adjacent moving surfaces by introducing a substance known as a lubricant between them. Lubricant facilitates the transmission of pressure or the load between the surfaces that are in contact. Lubricant operates as an anti-friction agent, enabling smooth operation, ensuring dependable machine performance, and reducing the chance of frequent malfunctions [1]. A lubricant can exist in the form of a liquid, a semi-solid substance like grease or a solid material [2]. The main purposes of lubrication are to reduce wear and decrease heat generation caused by the contact between moving surfaces, protect the surface from corrosion by limiting oxidation, and act as a barrier against dust, dirt, and water. Products made from fossil fuels, such as gasoline, plastics, and other chemicals, are crucial for industries and transportation systems around the world [3]. These goods have the potential to generate significant risks to the environment and pollution. In the realm of lubrication, with a significant use of petroleum products, it is possible to replace mineral-based lubricants with bio-lubricants. This substitution has the potential to mitigate certain environmental impacts associated with lubrication. Bio-lubricants exhibit superior lubricating characteristics in contrast to conventional lubricants, while also being renewable and biodegradable. Bio-lubricants are commonly derived from plant oils, polymeric polysaccharides, and wax esters [4].

A transesterification reaction is conducted during the production of bio-lubricants. This reaction entails the combination of an ester with an alcohol, leading to the formation of a different ester through the exchange of the alkyl group. Transesterification of vegetable oils (VOs) yields fatty acid alkyl esters with various lengths along the alcohol chain. The reaction is generally catalyzed by mineral acids and bases, and the final product, which is fatty acid alkyl esters, has the dormant to be used as lubricating agents, fuel, and biodiesel [5]. Any lubricant that breaks down naturally and poses no threat to humans or the environment is collectively referred

DOI: 10.1201/9781003717362-2

to as a bio-lubricant. They are renewable, produce no greenhouse gas emissions, and show promise as lubricants, although their use is currently far lower than that of mineral oil-based lubricants. The phrases mineral oil, base stock, and base oil are among the most important ones to learn [6]. Mineral oil is a liquid hydrocarbon by-product attained via the process of distilling crude oil. Irrespective of the raw material or the manufacturing site, base stock is generally produced by a single manufacturer according to the same standard (API 1509 2005). When producing lubricant, the base oil might consist of either a singular base stock or a blend of multiple base stocks [7].

Bio-lubricant is distinguished by its notable lubricating characteristics and its capacity to diminish frictional forces, wear, and friction coefficient when compared to other lubricants. Vegetable oils are considered highly effective lubricants because of their outstanding tribological characteristics. However, it has been observed that vegetable oils have lower thermal and hydrolytic stability compared to typical mineral oils. Therefore, it is necessary to improve these features [8].

2.2 THE ROLE, FUNCTIONS, AND NECESSITIES OF A LUBRICANT

A lubricant is a substance that reduces friction and wear between surfaces, allowing solid entities to move more easily in relation to each other. Lubricating oils provide other roles beyond their basic role of decreasing friction and wear [2]. These services include heat removal, corrosion prevention, and power transfer. In addition, lubricants must effectively offer a liquid barrier at the points where moving components come into contact and also remove any particles that result from wear. To carry out these operations, lubricating oils must have specific physical and chemical properties. The essential need for lubricants is that the oil must retain its fluidity across a broad spectrum of temperatures. Typically, the usable range of liquid is limited by the pour point (PP) at low temperatures and the flash point at high temperatures. To ensure that the lubricant can be pumped when the equipment is started at extremely low temperatures, the PP should be below a specific threshold [5]. A high flash point is required to guarantee safe operation and limit the evaporation at the highest operating temperature. To fulfill the needs of very demanding applications, such as lubricants for aircraft jet engines, it may be essential to have a liquid range that surpasses 3008 °C. The viscosity of the lubricant has a significant effect on its ability to reduce friction and wear, hence affecting its overall efficiency. Moreover, the biodegradability of a chemical is the paramount factor determining its environmental fate. Primary degradation refers to the initial stage of the decomposition of a chemical, where the original molecule ceases to exist.

Nevertheless, it is crucial to ascertain the degradability or mineralization of chemicals into CO_2, H_2O, and the production of biomass. The secure reintegration of the organic material in the natural carbon cycle is ensured by its ultimate biodegradability, which is vital for determining its environmental classification [9].

2.3 SOURCES AND PREPARATION OF LUBRICATING OILS

Traditional lubricants consist of complex hydrocarbons with a high molecular weight that are obtained from the residual oil left after the refining process. Bio-lubricants, now derived from vegetable oils, consist of unsaturated fatty acids as their primary components. These compounds are not suited for applications such as lubricating oils because they have low heat stability and poor oxidation stability due to the presence of oxidizable functional groups. Through the utilization of chemical modification methods such as transesterification, epoxidation, and hydrogenation, it is feasible to minimize the quantity of unsaturated fatty acids present in vegetable oils [10]. This modification makes the oils appropriate for use in engines, making them similar to traditional lubricants. Figure 2.1 depicts the life cycle of a bio-based lubricant.

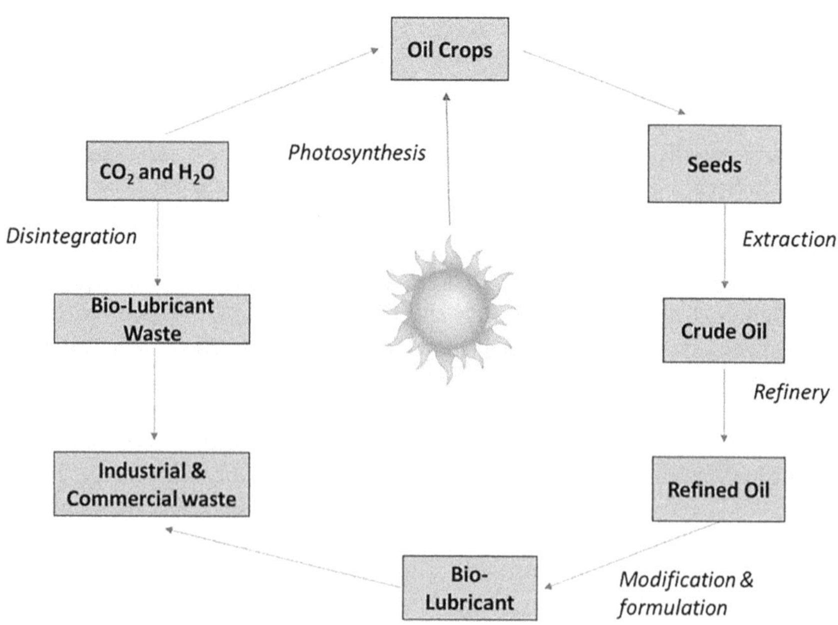

Figure 2.1 Life cycle of bio-based lubricant.

2.4 TYPES OF BIO-LUBRICANTS

2.4.1 Vegetable oils

Various seeds are used to produce vegetable oils. Throughout history, vegetable oils have been employed as lubricants. Images of different seeds, which are utilized for producing vegetable oils and have been found to possess beneficial lubricating characteristics in scientific investigations, have been documented and presented in Figure 2.2. Prior to the 19th century, they served as the main constituents of lubricating oils [11]. The swift process of industrialization led to an increase in the need for lubricants, resulting in challenges regarding the availability and cost of the lubricants generated from plants and animals. In the latter half of the 19th century, efficient research and extraction of mineral oils led to their widespread use as lubricants, replacing vegetable and animal oils. Mineral oils provided a variety of fluids with satisfactory lubricating properties, causing petroleum-based products to dominate the lubrication industry. However, the increased industrial use of mineral oils also resulted in numerous environmental problems. Interestingly, the inherent qualities of vegetable oil as a lubricant surpass certain characteristics that are found in other biomasses [12]. Mineral oil-based lubricants are unable to equal the lubricating characteristics of animal oils, vegetable oils, and fats. It is noteworthy that vegetable

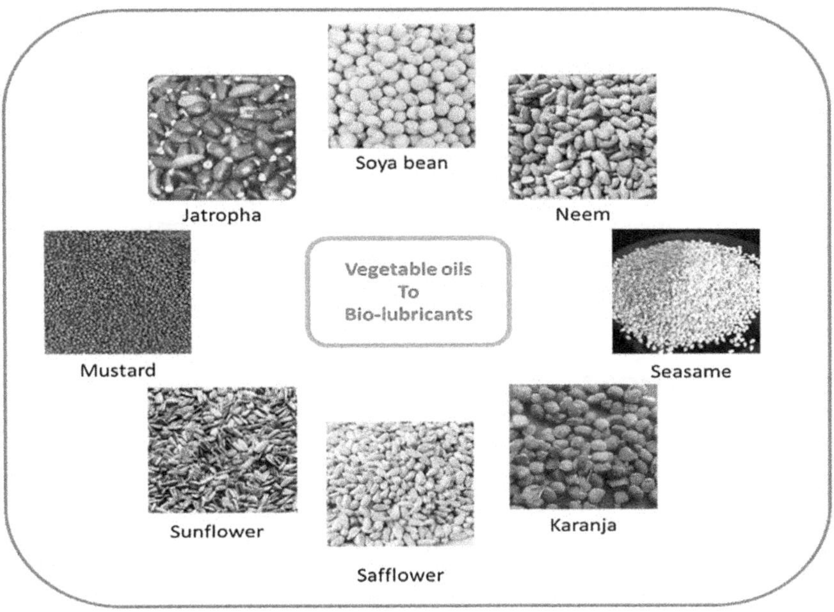

Figure 2.2 Commonly employed seeds for vegetable oil utilized in lubricating studies.

oils possess a remarkable capacity to stick to metal surfaces and generate a robust single layer, with the polar portion attaching to the surfaces and the hydrocarbon chains adopting different orientations. Moreover, vegetable oils are renowned for their fast biodegradability. As a result, there have been efforts to combine mineral oils with vegetable oils. A blend of mineral oil and mustard oil has been recognized as a highly promising lubricant [13].

Various varieties of vegetable oils have been recommended as main oils for bio-lubricants. An analysis is conducted to establish the tribological features of cottonseed oil as a lubricant for multi-cylinder engines, specifically focusing on its anti-wear capabilities [14]. It was suggested that cottonseed oil is not suited for use as a lubricant in its natural state because of its bigger wear scar width, even if it has a lower coefficient of friction than commercial lubricants. To enhance the anti-wear characteristics, one can either chemically alter the oil or use suitable anti-wear additives [15].

2.4.2 Synthetic oils

Synthetic esters are produced through the chemical alteration of oils derived from plants or fats obtained from animals. Currently, the majority of research on bio-lubricants has concentrated on enhancing their physicochemical characteristics using various chemical modification techniques and diverse catalytic procedures (Figure 2.3). Nevertheless, it is imperative to persist in the advancement of these bio-lubricants, since they still exhibit suboptimal thermal degradation, susceptibility to hydrolysis, and inadequate oxidative stability due to the presence of unsaturated vegetable oils [16]. The main challenge in improving the quality of bio-lubricants lies in the necessity to strengthen their resistance to oxidation and lower their pour temperature. The preferred techniques focus on modifying unsaturated oils by chemical processes, as double bonds are prone to reacting with oxygen in the surrounding environment. The most often used procedures for modifying vegetable oils include transesterification/esterification processes, epoxidation, hydrogenation, and estolide synthesis [17]. Generally, these methods improve specific physicochemical characteristics.

2.5 APPLICATIONS OF BIO-LUBRICANTS

Lubricants can be used in both enclosed and exposed systems. In open systems, lubricants are consistently discharged into the surroundings, whereas in closed systems, lubricants must be isolated to prevent unregulated discharges. Nevertheless, improper handling or human mistakes can lead to unintended disposal of lubricants into the surroundings. When evaluating potential replacements to standard mineral oils, it is important to prioritize lubricants with high levels of biodegradability [17, 18]. The significant characteristics of bio-lubricants for various applications are shown in Table 2.1.

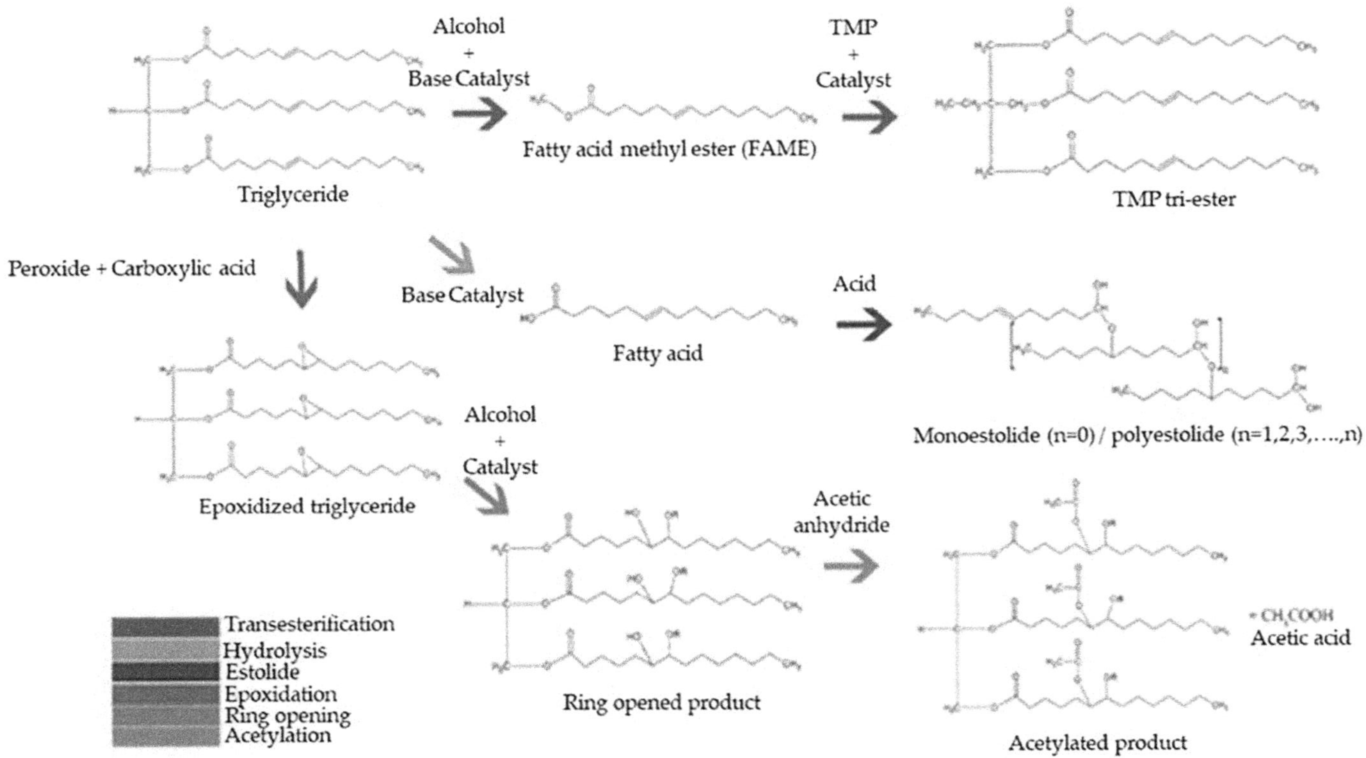

Figure 2.3 Chemical modifications of vegetable oils to obtain bio-lubricants.

Table 2.1 Key characteristics of bio-lubricants for various applications

Bio-lubricants	*Properties*	*Application*
Engine oil	Minimal emissions of volatile organic compounds and high lubricity.	Lowers engine emissions and increases engine performance.
Hydraulic oil	Weak compression and rapid air release.	Enhanced pressure transmission and lower vibration and noise.
Compressor oil	High thermal stability.	High temperature/pressure tolerance.
Metalworking oil	Less volatility, resistance to corrosion, excellent emulsification, and good lubricity.	Less harmful mist generation, tools last longer, emulsions stable at high temperature.
Transmission oil	Better lubricity and increased welding load	Additives that work can be included.
Chainsaw oil	Low volatility.	Less harmful mist generation.
Insulating oil	Higher level of water solubility and higher dielectric constant.	Reduces the moisture's impact on insulation, improved insulation.

2.6 INTRODUCTION OF ADDITIVES

Lubricant additives primarily serve to enhance the characteristics of the base stock in various working situations and meet the demanding performance standards of machines. Lubricant additives are chemical constituents that must effectively mix with the base oil in order to operate as a cohesive fluid. Additive manufacturers frequently market additive packages that consist of multiple additives blended together and diluted with a base oil at a greater concentration. The lubricant producer adds the additive package to the lubricant blend at a specific treat rate in order to achieve the desired performance [19]. The concentration of different additives is limited by considerations such as their main functions (e.g., dispersancy, wear protection, etc.), their synergistic or antagonistic interaction with other additives, and their industry restrictions. Lubricant additives can be grouped into different categories based on their overall functions of improving performance and extending service life. The first group consists of additives that enhance the lubricant by introducing additional properties, commonly referred to as surface protection additives. Some examples of additives are anti-wear additives, extreme pressure (EP) additives, corrosion inhibitors, detergents, and dispersants. Performance additives are a type of additive that improve the existing qualities of a lubricant. Viscosity index improvers, viscosity

Table 2.2 Chemical compounds used as additives with their objective [11]

Name of chemical	*Additive*	*Objective*
Polymethacrylates	Pourpoint depressant	Facilitate the flow of lubricant at low temperature.
Zinc dithiophosphate	Anti-wear additives	Lowered wear and friction.
Alkyl-substituted diphenyls	Antioxidants	Reduce oxidative degradation.
Succinic acid esters	Corrosion inhibitors	Prevent metal parts that come into contact with the lubricant from rusting and corrosion.
Carbodiimides	Hydrolysis protection	Prevent lubricant hydrolysis.
Deactivators	Benzotriazols	Decreases the lubricating reaction.

modifiers, friction modifiers, and pour point depressants are all the examples of this category. Lubricant protection additives, the third category of additives, are designed to mitigate the adverse effects or alterations that occur over the lifespan of the lubricant. These substances consist of antifoamants and antioxidants [20]. Table 2.2 displays the chemical substances utilized as additives, along with their respective purposes.

2.6.1 Vegetable oils as additives

A little quantity of either animal or vegetable oil has been incorporated into mineral oil-based lubricants as an additive to enhance their lubricating properties and give them a greasy texture. In order to meet the growing demand for lubricants across a range of applications, including high-speed internal combustion engines, numerous additives have been created specifically for mineral oil-based lubricants. These modifiers include inorganic, organic, and polymer substances. Compared to pure palm oil, the combination of palm oil and mineral engine oil showed a 25% improvement in wear, while the combination of soybean oil and mineral-based engine oil showed a 27% improvement in wear. An absence of a direct correlation between wear and friction was noted [21]. Soybean and palm oils demonstrated worse performance in comparison to conventional mineral oils, making them unsuitable for utilization as extreme contact engine oils or lubricity additives. In their study, Nagy et al. [22] employed vegetable oils, specifically sunflower, rapeseed, and castor oil, as additives to improve the characteristics of PLA/starch composites. The addition of a castor oil-based substance, produced by combining a solvent and heating it to a temperature of 140°C, resulted in a 20 MPa increase in the tensile strength of the test specimen containing 10% starch. In addition, the elongation at break of the specimen rose by 3.4% [22].

2.6.2 Additives from natural resources other than vegetable oils (VO)

Various natural compounds, such as microalgae oil, lignin, curcumin, chitosan, and cellulose, are utilized as natural additives in lubricants instead of vegetable oils. The red microalga *Dixoniella Grisea* has the ability to decrease friction in a tribo system. Antioxidant additives, including palmitoylated lignin and curcumin, were employed in VO/waste cooking oils. The lignin fractions extracted from straw and birch wood following palmitoylation showed superior antioxidant activity in comparison to butylated hydroxytoluene (BHT) when combined with castor oil [23]. The addition of azomethine-functionalized chitosan to paraffin base oil resulted in highly effective tribological performance. The combination of bio-phenols and 4-aminodiphenylamine by amidation results in the creation of extremely effective multifunctional additives for bio-based fluids such as rapeseed oil, coconut oil, and epoxy soybean oil [24]. They enhance the resistance of VOs to heat-induced oxidation, minimize friction, and prevent wear, regardless of the level of saturation. The four-ball test revealed that the additions resulted in a reduction of wear and friction by up to 18% and 25% in rapeseed oil and coconut oil, respectively.

2.6.3 Nanoparticles

The literature widely reports that nanoparticles are highly effective additives for lubricants and bio-based lubricants, as they effectively minimize friction and wear. Nanoparticles are considered preferable alternatives to traditional extreme pressure and anti-wear additives because of their favorable environmental characteristics [25]. The properties of bio-based lubricants are affected by the features of the nanoparticles, including their size, geometry, and concentration [26]. Several studies have explored the impact of TiO_2, CuO, and ZnO nanoparticles on the characteristics of bio-based lubricants [25–27]. According to Alves et al., the use of zinc oxide and copper oxide nanoparticles as anti-wear additives for soybean and sunflower oils was not effective [28]. In their study, Boshui et al. found that the inclusion of stearic acid-modified cerium borate nanoparticles to rapeseed oil improved its capability to decrease friction and wear, in contrast to the unmodified nanoparticles and base oil [29]. Gulzar et al. [30] conducted a study to investigate the effects of CuO and MoS_2 nanoparticles on enhancing the anti-wear properties of chemically functionalized palm oil.

2.7 BENEFITS OF LUBRICANT ADDITIVES

Lubricant additives bundles are outstanding items. The oils produced from these bundles are specifically formulated to offer optimal lubrication for any engine. In addition, oils can be classified for different purposes, such as for use in both gasoline and diesel engines [31]. The inclusion of polymeric consistency modifier additives in multigrade oils has eliminated the need for regular oil changes in operating conditions within atmospheric limits. These lubricants/greases offer the essential performance needed to meet the warranty demands set by the vehicle manufacturer, along with the protection of emission control systems [32]. Furthermore, it offers consumer assurance as well as protects the vehicle owner's investment. The integration of lubricant/grease additives in technology offers substantial benefits to clients in cost management related to vehicle design and operation. The expense associated with lubricant constitutes a minor proportion of the overall operational expenditure of the vehicle [33]. By integrating lubricant additives, consumers can cut their expenses by decreasing gasoline and oil consumption, minimizing repairs, prolonging service periods, lowering downtime losses, and improving vehicle durability. Lubricant additives bring significant benefits to the end consumer. Lubricant additives are vital components of modern oils. They are high-performance compounds that help keep engines and transmissions in better condition. A commitment to the continuing evolution of lubricant additive technology fosters the success of new engine designs that increase efficiency and conserve resources.

2.8 CONCLUSION

Modern machines operate under harsh conditions, demanding high-performance lubricants that are also environmentally friendly. Lubricant engineers rely on specially formulated additives to achieve these goals. This chapter explored the different types of lubricants, additives, and their functions. Lubricants are complex mixtures, and the interactions between different components can have a significant impact on performance. Developing effective lubricants requires extensive scientific knowledge and rigorous testing. As we move toward green tribology and industry 4.0, the future of lubricants will prioritize both performance and environmental responsibility.

REFERENCES

1. Robertson, W.S., 2019. Types and properties of lubricants. In W.L. Robertson ed., *Lubrication in Practice* (pp. 14–27). CRC Press.
2. Zhang, R., Liu, X., Guo, Z., Cai, M. and Shi, L., 2020. Effective sugar-derived organic gelator for three different types of lubricant oils to improve tribological performance. *Friction*, *8*, pp. 1025–1038.
3. Peng, Y., Wang, Z. and Zou, K., 2015. Friction and wear properties of different types of graphene nanosheets as effective solid lubricants. *Langmuir*, *31*(28), pp. 7782–7791.
4. Narayana Sarma, R. and Vinu, R., 2022. Current status and future prospects of biolubricants: Properties and applications. *Lubricants*, *10*(4), p. 70.

5. Syahir, A.Z., Zulkifli, N.W.M., Masjuki, H.H., Kalam, M.A., Alabdulkarem, A., Gulzar, M., Khuong, L.S. and Harith, M.H., 2017. A review on bio-based lubricants and their applications. *Journal of Cleaner Production, 168*, pp. 997–1016.
6. Salimon, J., Salih, N. and Yousif, E., 2010. Biolubricants: Raw materials, chemical modifications and environmental benefits. *European Journal of Lipid Science and Technology, 112*(5), pp. 519–530.
7. Panchal, T.M., Patel, A., Chauhan, D.D., Thomas, M. and Patel, J.V., 2017. A methodological review on bio-lubricants from vegetable oil based resources. *Renewable and Sustainable Energy Reviews, 70*, pp. 65–70.
8. Mobarak, H.M., Mohamad, E.N., Masjuki, H.H., Kalam, M.A., Al Mahmud, K.A.H., Habibullah, M. and Ashraful, A.M., 2014. The prospects of biolubricants as alternatives in automotive applications. *Renewable and Sustainable Energy Reviews, 33*, pp. 34–43.
9. Barbera, E., Hirayama, K., Maglinao, R.L., Davis, R.W. and Kumar, S., 2024. Recent developments in synthesizing biolubricants—A review. *Biomass Conversion and Biorefinery, 14*(3), pp. 2867–2887.
10. Ahmad, U., Naqvi, S.R., Ali, I., Naqvi, M., Asif, S., Bokhari, A., Juchelková, D. and Klemeš, J.J., 2022. A review on properties, challenges and commercial aspects of eco-friendly biolubricants productions. *Chemosphere, 309*, p. 136622.
11. Hamnas, A. and Unnikrishnan, G., 2023. Bio-lubricants from vegetable oils: Characterization, modifications, applications and challenges–Review. *Renewable and Sustainable Energy Reviews, 182*, p. 113413.
12. Huang, W., Tan, Y., Chen, B., Dong, J. and Wang, X., 2003. The binding of antiwear additives to iron surfaces: Quantum chemical calculations and tribological tests. *Tribology International, 36*(3), pp. 163–168.
13. Jabal, M.H. and Khalefa, M.Z., 2018. Tribological characteristics evaluation of mustard oil blends. *Journal of Engineering, 24*(3), pp. 1–12.
14. Aravind, A., Joy, M.L. and Nair, K.P., 2015. Lubricant properties of biodegradable rubber tree seed (Hevea brasiliensis Muell. Arg) oil. *Industrial Crops and Products, 74*, pp. 14–19.
15. Sammaiah, A., Padmaja, K.V. and Prasad, R.B.N., 2014. Synthesis of epoxy jatropha oil and its evaluation for lubricant properties. *Journal of Oleo Science, 63*(6), pp. 637–643.
16. Karmakar, G., Ghosh, P. and Sharma, B.K., 2017. Chemically modifying vegetable oils to prepare green lubricants. *Lubricants, 5*(4), p. 44.
17. Cecilia, J.A., Ballesteros Plata, D., Alves Saboya, R.M., Tavares de Luna, F.M., Cavalcante Jr, C.L. and Rodríguez-Castellón, E., 2020. An overview of the biolubricant production process: Challenges and future perspectives. *Processes, 8*(3), p. 257.
18. Willing, A., 2001. Lubricants based on renewable resources–An environmentally compatible alternative to mineral oil products. *Chemosphere, 43*(1), pp. 89–98.
19. Rudnick, L.R., 2009. *Lubricant Additives: Chemistry and Applications.* CRC Press.
20. Puhan, D., 2020. Lubricant and lubricant additives. In Amar Patnaik, Tej Singh and Vikas Kukshal eds., *Tribology in Materials and Manufacturing-Wear, Friction and Lubrication.* IntechOpen.

21. Bahari, A., Lewis, R. and Slatter, T., 2018. Friction and wear response of vegetable oils and their blends with mineral engine oil in a reciprocating sliding contact at severe contact conditions. *Proceedings of the Institution of Mechanical Engineers, Part J: Journal of Engineering Tribology*, *232*(3), pp. 244–258.
22. Nagy, B., Török, F., Tomasek, S. and Miskolczi, N., 2023. Vegetable oil based additives to enhance the properties of PLA/starch composites: The effect of reaction parameters. *Industrial Crops and Products*, *191*, p. 116025.
23. Rawat, S.S. and Harsha, A.P., 2021. Recent progress in vegetable oil-based lubricants for tribological applications. *Tribology and Sustainability*, CRC Press, pp. 115–142.
24. Samarth, N.B. and Mahanwar, P.A., 2015. Modified vegetable oil based additives as a future polymeric material. *Open Journal of Organic Polymer Materials*, *5*(1), p. 1.
25. Goyan, R.L., Melley, R.E., Wissner, P.A. and Ong, W.C., 1998. Biodegradable lubricants. *Tribology & Lubrication Technology*, *54*(7), p. 10.
26. Rhodes, B.N., Mammel, W., Landis, P. and Erickson, F.L., 1995. *Water rejection of vegetable oil base stocks for tractor/hydraulic fluids* (No. 952073). SAE Technical Paper.
27. Masripan, N.A., Salim, M.A., Omar, G., Mansor, M.R., Saad, A.M., Hamid, N.A., Syakir, M.I. and Dai, F., 2020. Vegetable oil as bio-lubricant and natural additive in lubrication: A review. *International Journal of Nanoelectronics & Materials*, *13*, 161–176.
28. Alves, S.M., Barros, B.S., Trajano, M.F., Ribeiro, K.S.B. and Moura, E.J.T.I., 2013. Tribological behavior of vegetable oil-based lubricants with nanoparticles of oxides in boundary lubrication conditions. *Tribology International*, *65*, pp. 28–36.
29. Boshui, C., Kecheng, G., Jianhua, F., Jiang, W., Jiu, W. and Nan, Z., 2015. Tribological characteristics of monodispersed cerium borate nanospheres in biodegradable rapeseed oil lubricant. *Applied Surface Science*, *353*, pp. 326–332.
30. Gulzar, M., Masjuki, H.H., Varman, M., Kalam, M.A., Mufti, R.A., Zulkifli, N.W.M., Yunus, R. and Zahid, R., 2015. Improving the AW/EP ability of chemically modified palm oil by adding CuO and MoS2 nanoparticles. *Tribology International*, *88*, pp. 271–279.
31. Mang, T. ed., 2014. *Encyclopedia of Lubricants and Lubrication* (Vol. 1). Berlin Heidelberg: Springer.
32. Srivyas, P.D. and Charoo, M.S., 2019. Effect of lubricants additive: Use and benefit. *Materials Today: Proceedings*, *18*, pp. 4773–4781.
33. Singh, A., Verma, N., Mamatha, T.G., Kumar, A., Singh, S. and Kumar, K., 2021. Properties, functions and applications of commonly used lubricant additives: A review. *Materials Today: Proceedings*, *44*, pp. 5018–5022.

Chapter 3

Tribology

A circular economy practice

Himanshu Chandra

3.1 INTRODUCTION

The coefficient of friction (CoF) is a unitless quantity; it is the ratio of friction to force and indicates the proportion between the frictional force acting between two surfaces and the force exerted on them in the direction perpendicular to the surfaces. Therefore, CoF is directly proportional to the friction when two surfaces are in contact and in relative motion, and the CoF is directly proportional to the amount of external force (energy) required to overcome friction. Therefore, the more the CoF, the higher the energy that has to be applied to the system. Thus, the efficiency of the system will decrease. However, when a lubricant is present, the CoF tends to decrease, which means there is less friction, and the motion between the surfaces becomes smoother; therefore, less energy is wasted on overcoming friction. Furthermore, a surface's lubricity enables it to move smoothly against another surface without causing damage, wear, or other problems.

Wear is the outcome of friction that results in any microscopic or macroscopic removal of surface material or deformation of any surface due to mechanical action. Friction and wear are the two sides of one coin and are related to one another. These terms, such as wear, CoF, and friction etc., come in handy when we are dealing with machines and mechanisms, as they all involve surface contact and the relative movement of at least two surfaces. Therefore, in order to comprehend the physical aspects related to the reliability of mechanical systems, it is imperative to possess a thorough comprehension of the engineering and scientific principles that provide the basis for their functioning. Tribology is a field of science that focuses on the intricate interaction of surfaces that are in motion relative to each other. Henceforth, tribology is a science that came into existence to minimise the role of friction that leads to: first, the wear that damages the material; second, the wastage of energy in overcoming frictional forces and thus decreases the efficiency of the system; and third, decreases the durability and longevity of the system.

The word "tribology" was first used in a landmark report by H.P. Jost in 1966. The term was coined specifically to represent all the phenomena

 DOI: 10.1201/9781003717362-3

related to friction and wear. It was derived from the Greek term *tribos*, meaning rubbing, so that the literal meaning would be "the science of rubbing" (Blau 2008). Though the term was coined not long ago, the concept behind this term has concerned various great minds since the era of Aristotle, dating back centuries (Hähner and Spencer 1998). Afterward, Leonardo Da Vinci developed two empirical friction laws. Leonardo da Vinci was the first to state the two laws of friction governing the motion of a rectangular block sliding over a flat surface. He discovered that the friction force is independent of the contact area and proportional to the normal load. In 1785, Coulomb introduced a third law, which states that the friction force is not dependent on sliding velocity (Scoones 2007). Since the beginning of the twentieth century, knowledge in all areas of tribology has expanded tremendously due to enormous industrial growth, which led to the demand for an improved understanding of tribology. Tribology as a concept is not only restricted to mechanics nowadays, but is also being studied and applied in various domains, such as in medical science, where tribology is being applied extensively in artificial joint replacements (Faivre et al. 2018; Milner et al. 2018; Zhang et al. 2018), in dental and oral science (Rivals et al. 2018; Santos et al. 2018; Uno et al. 2018; Sarkar et al. 2019), skin tribology focusing on friction and physical cognisance of skin (Liu et al. 2018; Vagts et al. 2018; Wu et al. 2018), ocular surface (Lievens and Rayborn 2022), etc., and various other domains as well. Since the time tribology came out as a concept, it has found its application in almost all the major domains.

The concept of tribology primarily originated to reduce the wastage of energy and improve the efficiency of a mechanism. Though the term tribology was coined in 1966, the concept of tribology came into existence when great thinkers, from Aristotle to Leonardo Da Vinci, realised that output is fairly less than the energy that is being externally fed into the mechanical system. Since there was a visible difference between input and output, they realised that there was a frictional force, which was going uncontrolled, decreasing the efficiency of the mechanism, and also resulting in the wear and tear of the components. Therefore, the primary objective behind the emergence of the concept of tribology was to control the contact forces being applied to the system, and tribology became more like a necessity to reduce the wastage of the energy that is being fed into the system and get the desired output with minimum energy input.

3.2 SUSTAINABILITY

Recently, "sustainability" has become a ubiquitous buzzword over the past two decades, permeating discussions on various aspects of human life and development. Initially coined by the German forester Hans Carl von Carlowitz in 1712 to prescribe long-term forest management, the concept gained widespread currency in the 1980s amid the burgeoning environmental movement and debates on the limits to growth. The pivotal moment occurred with the 1987 publication of the Brundtland Report, chaired by Gro Brundtland, which defined the sustainable development as meeting the present needs without compromising future generations' ability to meet their own. This definition marked the integration of environmental and economic concerns. The ensuing years witnessed a proliferation of academic and policy debates on sustainability, encompassing ecological, economic, and socio-political dimensions. The 1992 Rio conference emerged as a landmark, giving rise to high-level conventions on climate change, biodiversity, and desertification, along with local initiatives like Agenda 21 (Scoones 2007). While the 1990s witnessed a surge in planning approaches, measurement indicators, and projects under the umbrella of sustainability, the early 2000s brought critiques about the lack of progress and the bureaucratic nature of sustainability initiatives. Nevertheless, the concept persisted, evolving to address contemporary global challenges such as genetically modified crops, climate change, and pandemics, albeit facing criticism for perceived sell-outs to business interests. By the early 2000s, the sustainability movement had become more muted and fractured, but retained its symbolic and unifying role, with debates shifting towards resilience, diversity, and precaution. Despite challenges and changing terminologies, sustainability, as a guiding principle and normative commitment, seems poised to endure and adapt to the evolving landscape of global challenges.

3.3 CIRCULAR ECONOMY

The concept of the circular economy (CE) has received a lot of traction from the research community and is often viewed optimistically by various environmentalists and socialists. In the wake of an era where natural resources are being exploited extensively, the United Nations has accepted that it is high time to think of sustainability, and brought 17 sustainable development goals to protect the planet (United Nations 2023).

Though the concepts of CE and sustainability developed independently, various researchers have mentioned that CE can be used to achieve sustainable development (Homrich et al. 2018). The CE can be defined as a system that emphasises eliminating waste and emissions, limiting the use of raw resources, and minimising energy leakage (Geissdoerfer et al. 2017).

Since CE has been defined by various researchers and organisations throughout the year, the concept of CE can be understood by following four key properties: (a) retrieving resources and recycling post-use, (b) life extension, which can be achieved through improved design and regular maintenance, (c) minimising the consumption of resources, and (d) reduction of greenhouse gas (GHG) emissions. These four properties were outlined by Bocken et al. (2016). Therefore, any practice possessing these three properties can be termed a CE practice and thus promote sustainability.

3.4 TRIBOLOGY AND CIRCULAR ECONOMY

The relation between tribology and CE can be observed by examining tribology under four parameters mentioned below:

1. Minimising the consumption of resources

   Tribology is the science that focuses on controlling the friction between two moving parts. Since tribology ensures that the system does not overuse the energy and directly helps in reducing the consumption of one of the most important resources, that is, fuel and energy. Thus, tribology directly contributes to sustainability by improving energy efficiency (DeFries and Nagendra 2017; Ranius et al. 2018; OECD 2019; Shah et al. 2021; Ciulli 2022). This science of eliminating friction has resulted in contributing directly to the GDP of a country by reducing energy wastage. A study by Holmber and Erdemir (2019) showed that by improving tribological adaptation, a possible reduction of 40% energy wastage over the period of 15 years would actually produce 8% to 8.6% of annual savings in the GDP. Therefore, this aspect of tribology directly promotes sustainability.
2. Reducing GHG emissions

   The primary aim of CE and sustainability is to reduce GHG emissions, and various studies have shown that the adoption of tribology has been successful in reducing GHG emissions. Since tribology has proven to be essential in reducing the energy and fuel wastage, it also directly contributes to reducing CO_2 emissions. By reducing the friction losses to up to 40%, tribology saves more than one gigatons of GHG emissions per year (Holmberg and Erdemir 2019). It was reported that the majority of GHG emissions are caused by inefficient end-use energy, and the effective adoption of tribology can help control global warming by 2 degrees Celsius (Singh, Singh, and Sehgal 2016; Holmberg and Erdemir 2017). Thus, various researchers and studies have confirmed that the adoption of tribology in mechanical and moving systems will significantly reduce GHG emissions and improve economic profitability as well. Therefore, tribology directly contributes to sustainability by reducing GHG emissions.
3. Life extension

   Another aspect of CE is to increase the life span of the product and keep the product in use for a longer period of time. The major factor for the deterioration of a mechanical system is wear and tear caused by extensive, uncontrollable friction. Tribology aims to reduce friction by the adoption of various types of lubricants for specific material types. Therefore, effective adoption of tribology will reduce the impact of friction and thus reduce the wear and tear of the mechanical system. Since the wear and tear is significantly reduced, the mechanical system can operate for longer periods of time (Kalnes, Shonnard, and Schuppel 2006; Arimura, Hibiki, and Katayama 2008; Elango, Bornmann, and Kannan 2016; Anand et al. 2017; Kurdi et al. 2020). Thus, fulfilling the aspect of CE where the lifespan of products is extended and kept in the operating condition.

4. Retrieving resources and recycling post-use

   CE, as a concept, proposes that retrieving the resources post-use to further utilise them for either recycling, upcycling, or downcycling is very essential. Mostly, the resources are discarded before their intended use-life. Therefore, retrieving or extracting resources has been marked as one of the most important aspects of CE and promoting sustainability. The practice of extracting lubricants and further utilising them as auxiliary fuels has been followed for years (Al-Omari 2008; Arpa, Yumrutas, and Alma 2010; Lam, Russell, and Chase 2010). Various studies are exploring how to improve the properties of lubricants post-use to reuse them. Many studies have suggested that the presence of chemical additives can improve the usability of lubricants post-use and extend the lifespan of lubricants (Liu et al. 2013; Pereira et al. 2017).

3.5 CONCLUSION

Tribology is an old science developed for controlling the friction between two moving joints and has proved to be one of the most essential theoretical and practical domains of the century. Friction, which creates resistance between two moving surfaces, causes numerous problems such as wear and tear, increased consumption of energy, and reduction in the expected life of the product, etc., which makes it undesirable. However, tribology has been successful in managing all the negative effects of friction and has helped in overcoming various aspects such as energy efficiency, GHG emissions, economic benefits, waste reduction, and recycling etc. Previous literature has tried to form a relationship between tribology and sustainability. However, the focus has been away from the fact that both concepts are the same part of the umbrella and achieve the same objectives, which are beneficial to the planet and to society. Therefore, in this study, based on four essential CE parameters, we have shown that tribology qualifies as a CE practice that promotes sustainability. We hope that in the future, a more focused study could be performed that dwells on specialised aspects of tribology, such as designing, lubrication, and material properties, etc., which may form a stronger bond of tribology with CE.

REFERENCES

Al-Omari, Salah B. 2008. "Used Engine Lubrication Oil as a Renewable Supplementary Fuel for Furnaces." *Energy Conversion and Management* 49 (12): 3648–53. https://doi.org/10.1016/j.enconman.2008.07.011

Anand, Ankush, Mir Irfan Ul Haq, Karan Vohra, Ankush Raina, and Mohammad Farooq Wani. 2017. "Role of Green Tribology in Sustainability of Mechanical Systems: A State of the Art Survey." *Materials Today: Proceedings* 4 (2): 3659–65. https://doi.org/10.1016/j.matpr.2017.02.259

Arimura, Toshi H., Akira Hibiki, and Hajime Katayama. 2008. "Is a Voluntary Approach an Effective Environmental Policy Instrument?" *Journal of Environmental Economics and Management* 55 (3): 281–95. https://doi.org/10.1016/j.jeem.2007.09.002

Arpa, O., R. Yumrutas, and M.H. Alma. 2010. "Effects of Turpentine and Gasoline-like Fuel Obtained from Waste Lubrication Oil on Engine Performance and Exhaust Emission." *Energy* 35 (9): 3603–13. https://doi.org/10.1016/j.energy.2010.04.050

Blau, Peter J. 2008. *Friction Science and Technology*. CRC Press. https://doi.org/10.1201/9781420054101

Bocken, Nancy M.P., Ingrid De Pauw, Conny Bakker, and Bram Van Der Grinten. 2016. "Product Design and Business Model Strategies for a Circular Economy." *Journal of Industrial and Production Engineering* 33 (5): 308–20. https://doi.org/10.1080/21681015.2016.1172124

Ciulli, Enrico. 2022. "Tribology and Sustainable Development Goals." In *Proceedings of I4SDG Workshop 2021*, edited by Giuseppe Quaglia, Alessandro Gasparetto, Victor Petuya, and Giuseppe Carbone, 108: 438–47. Mechanisms and Machine Science. Cham: Springer International Publishing. https://doi.org/10.1007/978-3-030-87383-7_48

DeFries, Ruth, and Harini Nagendra. 2017. "Ecosystem Management as a Wicked Problem." *Science* 356 (6335): 265–70. https://doi.org/10.1126/science.aal1950

Elango, Bakthavachalam, Lutz Bornmann, and Govindaraju Kannan. 2016. "Detecting the Historical Roots of Tribology Research: A Bibliometric Analysis." *Scientometrics* 107 (1): 305–13. https://doi.org/10.1007/s11192-016-1877-6

Faivre, Jimmy, Guillaume Sudre, Alexandra Montembault, Stéphane Benayoun, Xavier Banquy, Thierry Delair, and Laurent David. 2018. "Bioinspired Microstructures of Chitosan Hydrogel Provide Enhanced Wear Protection." *Soft Matter* 14 (11): 2068–76. https://doi.org/10.1039/C8SM00215K

Geissdoerfer, Martin, Paulo Savaget, Nancy M.P. Bocken, and Erik Jan Hultink. 2017. "The Circular Economy – A New Sustainability Paradigm?" *Journal of Cleaner Production* 143 (February): 757–68. https://doi.org/10.1016/j.jclepro.2016.12.048

Hähner, Georg, and Nicholas Spencer. 1998. "Rubbing and Scrubbing." *Physics Today* 51 (9): 22–27. https://doi.org/10.1063/1.882435

Holmberg, Kenneth, and Ali Erdemir. 2017. "Influence of Tribology on Global Energy Consumption, Costs and Emissions." *Friction* 5 (3): 263–84. https://doi.org/10.1007/s40544-017-0183-5

Holmberg, Kenneth, and Ali Erdemir. 2019. "The Impact of Tribology on Energy Use and CO2 Emission Globally and in Combustion Engine and Electric Cars." *Tribology International* 135 (July): 389–96. https://doi.org/10.1016/j.triboint.2019.03.024

Homrich, Aline Sacchi, Graziela Galvão, Lorena Gamboa Abadia, and Marly M. Carvalho. 2018. "The Circular Economy Umbrella: Trends and Gaps on Integrating Pathways." *Journal of Cleaner Production* 175 (February): 525–43. https://doi.org/10.1016/j.jclepro.2017.11.064

Kalnes, Tom N., David R. Shonnard, and Andreas Schuppel. 2006. "LCA of a Spent Lube Oil Re-Refining Process." In *Computer Aided Chemical Engineering*, 21: 713–18. Elsevier. https://doi.org/10.1016/S1570-7946(06)80129-X

Kurdi, Abdulaziz, Nahla Alhazmi, Hatem Alhazmi, and Thamer Tabbakh. 2020. "Practice of Simulation and Life Cycle Assessment in Tribology—A Review." *Materials* 13 (16): 3489. https://doi.org/10.3390/ma13163489

Lam, Su Shiung, Alan D. Russell, and Howard A. Chase. 2010. "Microwave Pyrolysis, a Novel Process for Recycling Waste Automotive Engine Oil." *Energy* 35 (7): 2985–91. https://doi.org/10.1016/j.energy.2010.03.033

Lievens, Chris W., and Elyse Rayborn. 2022. "Tribology and the Ocular Surface." *Clinical Ophthalmology* 16 (March): 973–80. https://doi.org/10.2147/OPTH.S360293

Liu, JianFang, KaLi Gu, HaiTao Duan, Yuan Zhao, and Jian Li. 2013. "Tribological and Economic Evaluation of Recycled Mineral Lubricating Oils." *Science China Technological Sciences* 56 (12): 2964–72. https://doi.org/10.1007/s11431-013-5408-x

Liu, X., M.J. Carré, Q. Zhang, Z. Lu, S.J. Matcher, and R. Lewis. 2018. "Measuring Contact Area in a Sliding Human Finger-pad Contact." *Skin Research and Technology* 24 (1): 31–44. https://doi.org/10.1111/srt.12387

Milner, Piers E., Maria Parkes, Jennifer L. Puetzer, Robert Chapman, Molly M. Stevens, Philippa Cann, and Jonathan R.T. Jeffers. 2018. "A Low Friction, Biphasic and Boundary Lubricating Hydrogel for Cartilage Replacement." *Acta Biomaterialia* 65 (January): 102–11. https://doi.org/10.1016/j.actbio.2017.11.002

OECD. 2019. *Global Material Resources Outlook to 2060: Economic Drivers and Environmental Consequences*. OECD. https://doi.org/10.1787/9789264307452-en

Pereira, O., J.E. Martín-Alfonso, A. Rodríguez, A. Calleja, A. Fernández-Valdivielso, and L.N. López De Lacalle. 2017. "Sustainability Analysis of Lubricant Oils for Minimum Quantity Lubrication Based on Their Tribo-Rheological Performance." *Journal of Cleaner Production* 164 (October): 1419–29. https://doi.org/10.1016/j.jclepro.2017.07.078

Ranius, Thomas, Aino Hämäläinen, Gustaf Egnell, Bengt Olsson, Karin Eklöf, Johan Stendahl, Jörgen Rudolphi, Anna Sténs, and Adam Felton. 2018. "The Effects of Logging Residue Extraction for Energy on Ecosystem Services and Biodiversity: A Synthesis." *Journal of Environmental Management* 209 (March): 409–25. https://doi.org/10.1016/j.jenvman.2017.12.048

Rivals, Florent, Kevin T. Uno, Faysal Bibi, Michael C. Pante, Jackson Njau, and Ignacio De La Torre. 2018. "Dietary Traits of the Ungulates from the HWK EE Site at Olduvai Gorge (Tanzania): Diachronic Changes and Seasonality." *Journal of Human Evolution* 120 (July): 203–14. https://doi.org/10.1016/j.jhevol.2017.08.011

Santos, F., A. Branco, M. Polido, A.P. Serro, and C.G. Figueiredo-Pina. 2018. "Comparative Study of the Wear of the Pair Human Teeth/Vita Enamic® vs Commonly Used Dental Ceramics through Chewing Simulation." *Journal of the Mechanical Behavior of Biomedical Materials* 88 (December): 251–60. https://doi.org/10.1016/j.jmbbm.2018.08.029

Sarkar, Anwesha, Efren Andablo-Reyes, Michael Bryant, Duncan Dowson, and Anne Neville. 2019. “Lubrication of Soft Oral Surfaces.” *Current Opinion in Colloid & Interface Science* 39 (February): 61–75. https://doi.org/10.1016/j.cocis.2019.01.008

Scoones, Ian. 2007. “Sustainability.” *Development in Practice* 17 (4–5): 589–96. https://doi.org/10.1080/09614520701469609

Shah, Raj, Mathias Woydt, Nabill Huq, and Andreas Rosenkranz. 2021. “Tribology Meets Sustainability.” *Industrial Lubrication and Tribology* 73 (3): 430–35. https://doi.org/10.1108/ILT-09-2020-0356

Singh, Sanjeev Kumar, Shyam Singh, and Ajay Kumar Sehgal. 2016. “Impact of Low Viscosity Engine Oil on Performance, Fuel Economy and Emissions of Light Duty Diesel Engine.” In *SAE 2016 International Powertrains, Fuels & Lubricants Meeting*, 2016-01–2316. https://doi.org/10.4271/2016-01-2316

United Nations. 2023. “The Sustainable Development Goals Report 2023.” Special Edition. https://unstats.un.org/sdgs/report/2023/

Uno, Kevin T., Florent Rivals, Faysal Bibi, Michael Pante, Jackson Njau, and Ignacio De La Torre. 2018. “Large Mammal Diets and Paleoecology across the Oldowan–Acheulean Transition at Olduvai Gorge, Tanzania from Stable Isotope and Tooth Wear Analyses.” *Journal of Human Evolution* 120 (July): 76–91. https://doi.org/10.1016/j.jhevol.2018.01.002

Vagts, Steffen, Josef Schlattmann, Alexander Kovalev, and Stanislav Gorb. 2018. “The Topology of the Leg Joints of the Beetle Pachnoda Marginata (Scarabaeidae, Cetoniinae) and Its Implication for the Tribological Properties.” *Biomimetics* 3 (2): 12. https://doi.org/10.3390/biomimetics3020012

Wu, Weibin, Christian Lutz, Simon Mersch, Richard Thelen, Christian Greiner, Guillaume Gomard, and Hendrik Hölscher. 2018. “Characterization of the Microscopic Tribological Properties of Sandfish (*Scincus Scincus*) Scales by Atomic Force Microscopy.” *Beilstein Journal of Nanotechnology* 9 (October): 2618–27. https://doi.org/10.3762/bjnano.9.243

Zhang, Xuewei, Jian Wang, Hui Jin, Shutao Wang, and Wenlong Song. 2018. “Bioinspired Supramolecular Lubricating Hydrogel Induced by Shear Force.” *Journal of the American Chemical Society* 140 (9): 3186–89. https://doi.org/10.1021/jacs.7b12886

Chapter 4

Tribological behavior of lithium-based coconut grease with graphene nanoplatelets

Shariq Ahmad Bhat, S. Aadil Hussain, and M. S. Charoo

4.1 INTRODUCTION

The demand for alternative, ecologically acceptable materials to replace crude mineral oils has sparked a rising interest in bio-based lubricants among industrial researchers. The initiatives are primarily motivated by the depletion of mineral oil deposits and concerns about environmental safety. Bio-based lubricants are made mostly from living elements, such as plant and animal fats. Bio-based lubricants are not only ecologically friendly, but are also renewable, biodegradable, lucrative, nontoxic, and non-accumulative [1]. Worldwide, vegetable oils, both edible and inedible, have been shown to be efficient bio-lubricants. This is in line with the favorable rise of worldwide vegetable oil output as a result of agricultural progress [2]. Vegetable oils are the ideal alternative/replacement to mineral oil by virtue of their triglyceride structure, high viscosity, viscosity index, and flash point [3]. Previous investigations have shown that the good lubricating characteristics may be achieved without or with the use of specialized additives [4–7].

Grease is typically utilized in components where the lubricant demands last longer without leaking or draining because of centrifugal or gravitational force. In accordance with Kumar and Nagendramma [8], grease's biodegradability is reflected by base oil, which is a major element that accounts for at least 80% of the overall composition. However, Ortansa et al. [9] explored the effects of thickeners and additives on the biodegradability of grease. Recent researches have examined the formulation and development of biogrease with an emphasis on tribological and physicochemical properties, utilizing vegetable oils as the base oil. Adhvaryu et al. [10] claim that creating lubricating grease is a difficult process that requires trial-and-error, where achieving the right grease consistency depends heavily on reactant and reaction process optimization. When creating metal soap grease, the proper ratios of base oil and thickener must be employed to ensure that the grease has the required consistency and thermal stability.

In addition to creating karanja oil-based lithium grease, Panchal et al. [11] investigated the efficacy of several fatty acid esters that had been

 DOI: 10.1201/9781003717362-4

chemically changed with hexanol, octanol, and neopentyl glycol. The formulated grease significantly outperforms mineral-based grease in terms of physicochemical and tribological properties, according to the study.

Padgurskas et al. [12] evaluated the implementation of lard grease with a sodium base and rapeseed grease with a lithium base suspended in ethanol. This study found that adding additives that create a wear-protective layer on contact surfaces increases the wear resistance of rapeseed oil greases by 40–50%. In greases that contain lard, however, the addition's effectiveness is reduced.

The additives used in lubricants are chosen based on the attributes that need to be improved. With the exception of solid additives, which are primarily employed in greases to boost load-bearing ability, the additives used in oil and grease are equivalent. Nanoparticles have recently gained popularity as a new type of additive [13]. Nanoparticles are used in a variety of fields, including industrial lubricants [14]. Nanoparticles exhibit unique lubricating and tribological characteristics. Nanoparticles can extend the life of lubricating oil and, as a result, the life of bearings in machines, saving millions of dollars in emergency repair. Nanoparticles when added to a lubricant reduce wear and friction, as reported in a number of studies [15–20].

As a lubricant additive, graphene has demonstrated excellent tribological properties. Graphene, a 2D material, has a large surface area, little interlayer linkage, and excellent chemical stability. The production of nano scrolls at the gliding interface gives it the potential to create super lubricity at the tribo-pair [21]. Because graphene may form a nano-bearing sheet, a small amount of graphene added to the lubricant can reduce the friction coefficient [22]. The lubrication procedure of graphene involves the formation of a protective film, interlayer sliding, and a synergetic effect [23–26]. Omrani et al. [27] observed significant reductions in friction and wear when GnPs were incorporated into canola oil. La et al. [28] discovered that adding 0.05 wt.% of GnPs enhances the tribological characteristics of engine oil significantly and reduces wear scarring by 35%. Suresha et al. [29] discovered that adding 1% GnPs to neem oil curtails WSD and CoF by 21% and 23%, respectively. To our knowledge, there have not been many references in the scientific literature discussing the effect of GnPs as an additive on the tribological behavior of coconut grease based on lithium.

Based on lithium, this study investigates the tribological characteristics of coconut fat. The authors claim that no studies on the ramifications of GnP on the tribological characteristics of coconut grease have been conducted. Due to its high oxidation and heat stability, coconut oil is used as the base oil in this investigation. Lauric acid is one of the saturated fatty acids that make up 92% of coconut oil [30, 31]. The oil's iodine value is 6–10, which indicates how unsaturated it is [30]. In addition, coconut oil has excellent lubricity, viscosity, and flash points. Because lithium-based grease has a high-temperature tolerance, it is ideal for heavy-duty lubrication and offers resistance to rust and corrosion; lithium stearate is used as a thickener. Graphene nanoplatelets find its application as an additive because of their remarkable tribological characteristics, substantial surface area, minimal interlayer linkage, and exceptional chemical stability. The sp2-bonded carbon two-dimensional lattice of GnPs enhances the material's thermal performance by facilitating effective heat transmission. The aim of this investigation is to infer the tribological behavior of grease based on coconut oil that has been added with various GnP concentrations.

4.2 EXPERIMENTAL PROCEDURE

4.2.1 Materials

Lithium stearate (thickener), graphene nanoplatelets (additive), and coconut oil (base oil) are the initial ingredients used to produce nano-biogrease. GnPs with a thickness of 2–4 nm are purchased from Nanoshel, India, and Table 4.1 lists their attributes. Three distinct weight percentages (0.5, 1, and 2 wt.%) are combined with the GnPs. We buy lithium stearate and coconut oil from a reliable local Indian provider.

4.2.2 Machine

A four-ball tribometer is used to calculate the characteristics of nano-grease in accordance with ASTM D-2266, as shown in Figure 4.1. A special collet holds the top ball, whereas the ball pot holds the three lower balls. The three lower balls are forced against a top ball. According to the test criteria, the top ball rotates at a specific speed. The grease sample is heated to 75 °C in the ball pot (oil cup) and then driven upward with a force against the top ball. After that, the top ball is driven at 1200 rpm by the drive motor for 60 minutes. The lower balls get scarred when the driving spindle rotates, causing frictional torque. The test lubricant is then poured from the ball pot, and the scar region is cleaned with a tissue. The steel ball surfaces' wear is measured by the diameter of the wear scar.

Table 4.1 Properties of graphene nanoplatelets

Appearance	*Black powder*
Density	2.3 g/cm^3
Morphology	Platelets
Thermal conductivity	3000 W/MK
Melting point	3652 °C

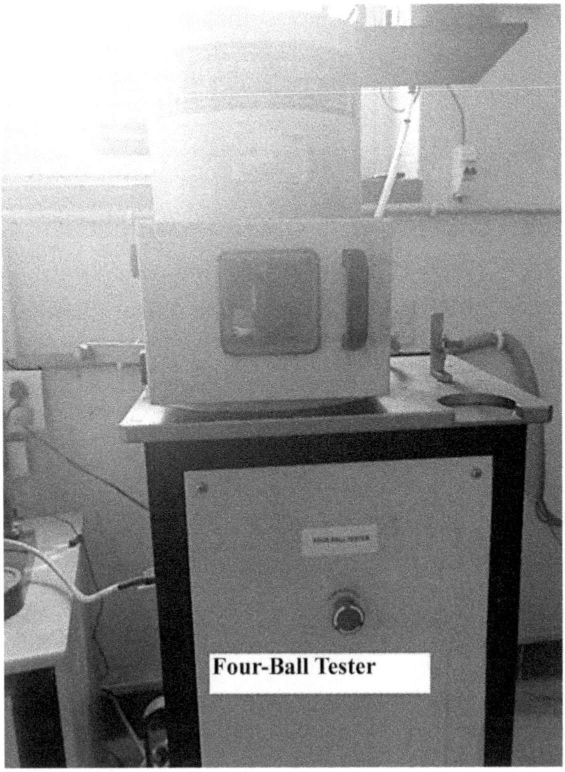

Figure 4.1 Four ball tester.

4.2.3 Nano-coconut grease formulation

To make nano-biogrease, the nanoparticles are added to the required amount of coconut oil in concentrations of 0.5, 1, and 2 wt.%. The oil-nanoparticle solution is sonicated for around 4 hours with a bath sonicator to achieve a homogeneous composition. The solution is then heated to 100 °C using a hot plate magnetic stirrer. Subsequently, the oil-nanoparticle solution is mixed with lithium stearate at the appropriate proportion (16 wt.%). When the oil, nanoparticle, and thickener mixture is heated to 210 °C, the oil starts to behave like grease. The mixture is held at 210 °C for 20 minutes and then progressively cooled, resulting in the formation of nano-grease.

4.2.4 Friction and wear tests

The tribological performance of graphene nanoplatelets as an additive on coconut grease is examined on a four-ball tester according to ASTM-D2266. At loads of 100, 200, 300, 400, and 500 N, the mean Hertz pressures are calculated to be 2.89, 3.51, 4.03, 4.41, and 4.8 GPa, respectively. Using an optical microscope, the diameter of the wear scars on the bottom balls is measured. For every lubricant sample, the tests are run three times, and a computer connected to a four-ball tester displays the friction coefficients. The computer is programmed to automatically calculate the CoF.

4.3 RESULTS AND DISCUSSION

4.3.1 Characterization of graphene nanoplatelets

The X-ray diffraction (XRD) spectrum of graphene nanoplatelets is depicted in Figure 4.2. The XRD study is performed at a scan speed of 8.2551°/minute and a scan range of 10–100°, respectively. The structure of GnP particles is confirmed by XRD with alpha = 90.3°, beta = 90°, and gamma = 120°. A DB card number of 01–082-9929 and a crystalline size of 56 Å are obtained. The crystallinity and purity are confirmed by the XRD spectrum.

4.3.2 Tribology of graphene nanoplatelets

Figure 4.3 shows how CoF changes as a function of GnP concentration. It has been discovered that the CoF of coconut fat containing GnPs steadily drops when GnP concentrations vary from 0.5 to 2 wt.%. In other words,

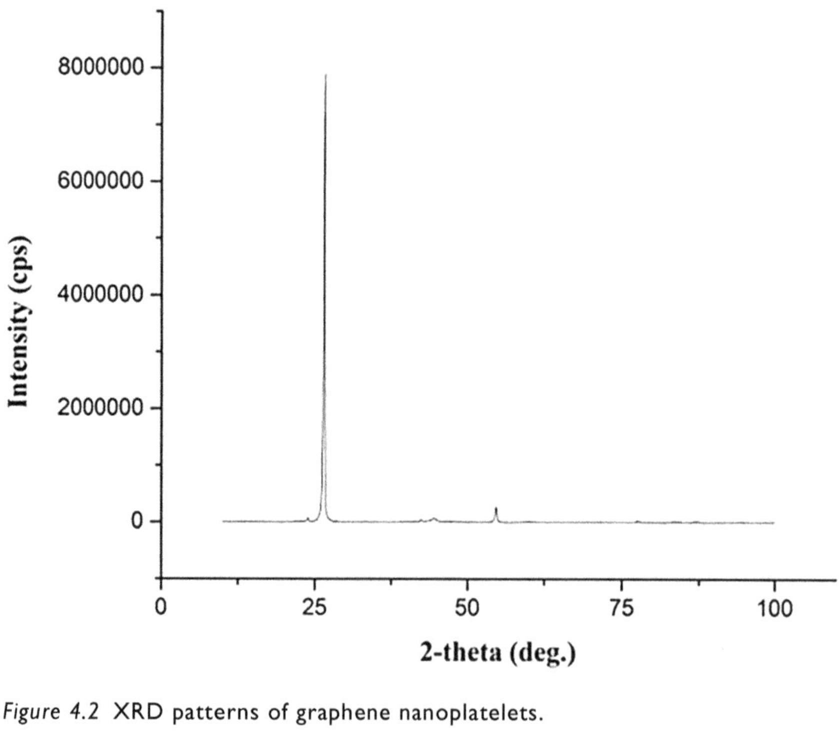

Figure 4.2 XRD patterns of graphene nanoplatelets.

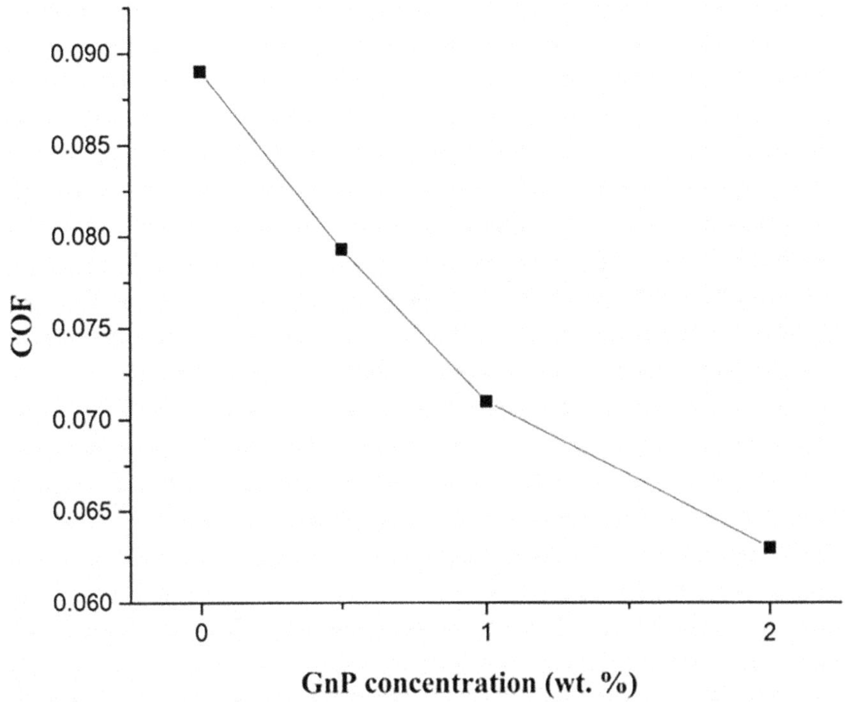

Figure 4.3 Variation of COF of coconut grease with GnP concentration.

GnPs may improve the properties of coconut grease that minimize the friction, and the ideal GnP concentration is 2 wt.%, where the CoF decreases by 29.21%.

The effect of GnPs on the wear scar diameter is shown in Figure 4.4 that shows the result of coconut grease, and Figure 4.5 shows the wear scar photographs. The wear characteristics of coconut grease are significantly impacted by the addition of graphene nanoplatelets. Figure 4.4 shows that the most efficient weight percentage of GnPs to reduce wear is 2 wt.%. At a 2 wt.% concentration of GnPs, the lowest WSD measured is 658 μm. Figure 4.6 shows the percentage increase in wear characteristics with the addition of GnPs.

Figure 4.7 depicts the variation of CoF with load for base coconut grease and coconut grease containing 2 wt.% GnPs. It is perceived that the CoF of lithium-based coconut grease containing 2 wt.% GnPs are lower than the base coconut grease at all loads. However, as the load increases, the CoF of both base coconut oil and oil with 2 wt.% GnPs increases. Consequently, the findings indicate that adding GnPs to lithium-based coconut oil improves its ability to reduce friction.

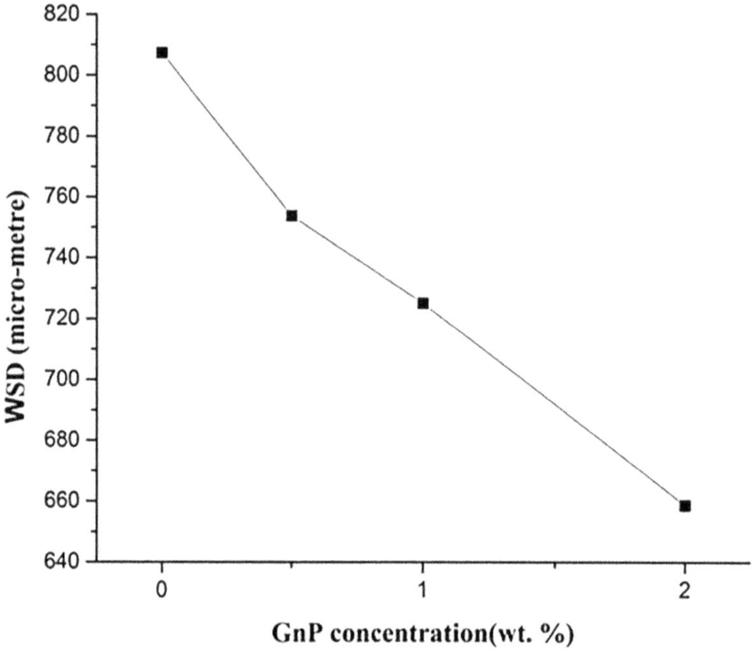

Figure 4.4 Effect of GnPs concentration on wear scar diameter of coconut grease.

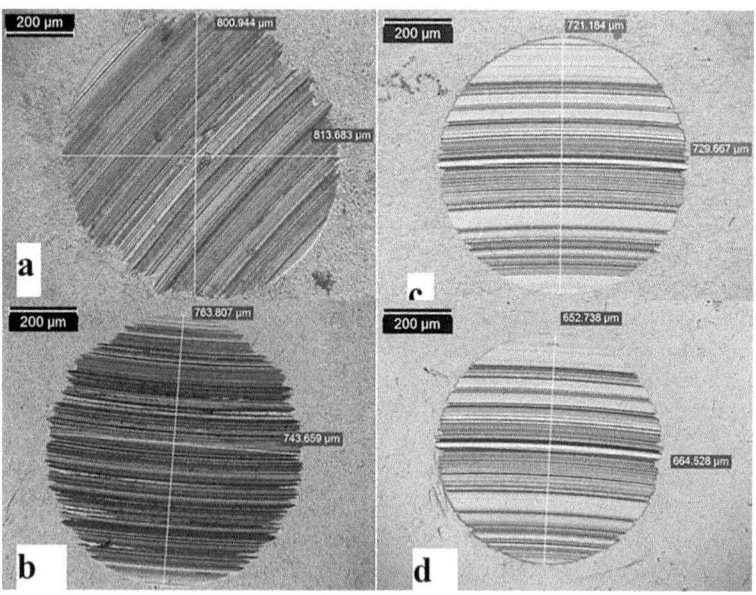

Figure 4.5 Optical images of wear scar of balls lubricated by (a) lithium based coconut grease (b) lithium based coconut grease + 0.5 wt.% GnPs (c) lithium based coconut grease + 1 wt.% GnPs (d) lithium based coconut grease + 2 wt.% GnPs.

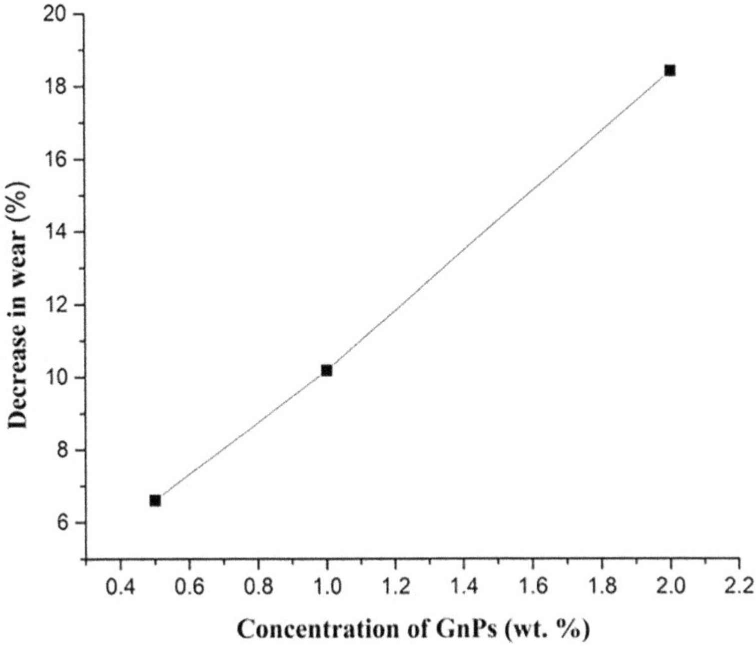

Figure 4.6 Percentage improvements in WSD with the incorporation of GnPs.

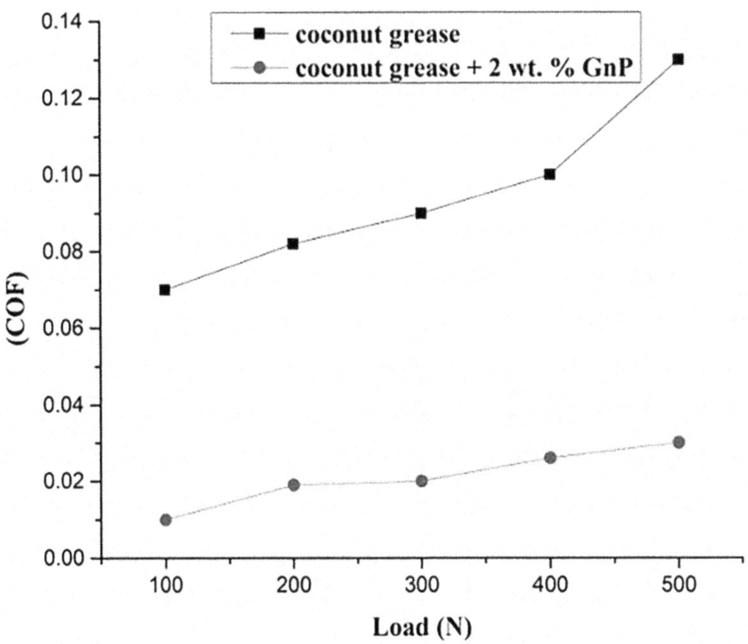

Figure 4.7 Variation of COF with load for base coconut grease and coconut grease containing 2 wt.% GnPs.

4.3.3 FESEM of worn surface

Field emission scanning electron microscopy (FESEM) is used to analyze the worn surface. The images of worn steel ball surfaces treated with coconut oil and coconut grease with 2 wt.% GnPs are displayed in Figure 4.8. Figure 4.8(a) reveals that the surface lubricated only with coconut grease has a coarser texture and, in some places, has mild adherence. Conversely, as demonstrated in Figure 4.8(b), adhesion is greatly impeded on the surface treated with coconut grease containing 2 wt.% GnPs. In addition, compared to balls lubricated with coconut oil without GnPs, the wear scar on balls treated with coconut grease containing 2 wt.% GnPs is significantly smoother and smaller. This demonstrates that the coconut grease containing 2 wt.% GnPs has good anti-wear properties.

4.3.4 Energy dispersive X-ray of worn surface

The energy-dispersive X-ray (EDX) spectra obtained from the worn scar treated with coconut grease based on lithium and coconut grease containing 2 wt.% GnPs are displayed in Figure 4.9. The worn scar surface is found to contain the element carbon (C), with an atomic concentration of 42.9% in coconut grease (which contains 2% weight percentage GnPs) and 29.2% in base grease. This demonstrates how the action of friction causes carbon in GnPs to collect on the worn surface of the steel ball under test. The presence of C provides strong evidence that the appearance of GnPs, which prevented the steel from steel contact, must have resulted in the establishment of a lubricating layer.

Figure 4.8 FESEM Images of surfaces of worn steel balls lubricated with (a) lithium based coconut grease and (b) lithium based coconut grease with 2 wt.% GnPs.

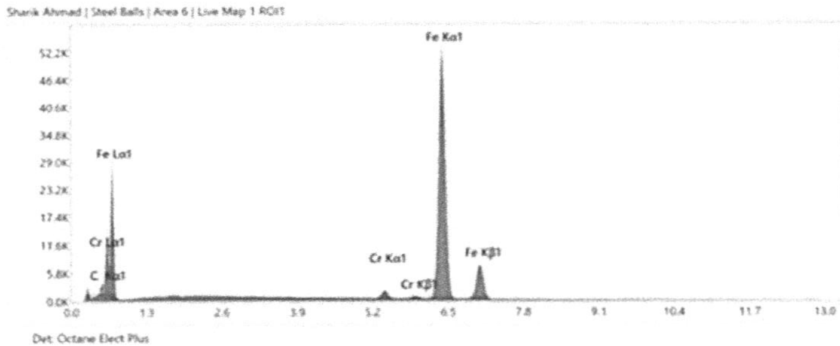

Element	Weight %	MDL	Atomic %
C K	8.2	0.50	29.2
Cr K	1.3	0.06	1.1
Fe K	90.5	0.11	69.7

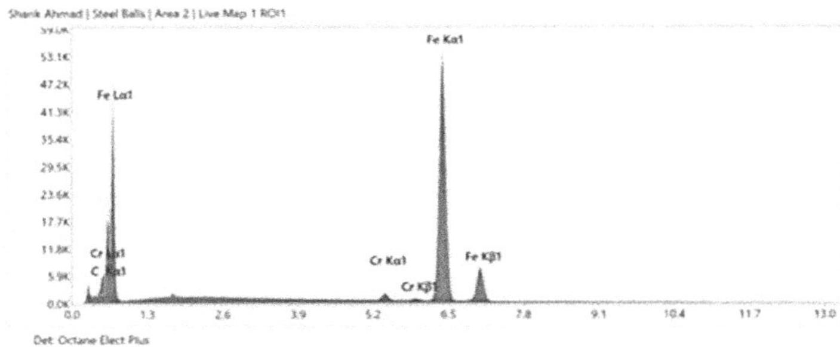

Element	Weight %	MDL	Atomic %
C K	13.9	0.46	42.9
Cr K	1.1	0.06	0.8
Fe K	85.0	0.13	56.3

Figure 4.9 EDX acquired from the worn scar lubricated by coconut grease and the coconut grease with 2wt.% GnPs.

4.4 CONCLUSION

Tribological applications are served by the fabrication of GnP composite nano-biogrease. Tests are conducted to determine how well the produced grease improves the tribological performance at various GnP concentrations. When added to coconut grease, GnPs significantly improve the grease's ability to withstand wear, reduce friction, and support loads. At 2 wt.% GnPs, the wear scar diameter decreases approximately by 18.41% and the friction coefficient by approximately 29.21%. The coconut grease with 2 wt.% GnPs exhibits favorable outcomes. According to the EDX study, there is a 42.9% atomic concentration of carbon on the worn scar surface. The carbon content indicates that GnPs led to the initiation of a lubricating layer, which reduced the amount of steel-on-steel contact.

REFERENCES

[1] Sharma, U. C., and N. E. E. T. U. Singh. "Biogreases for environment friendly lubrication." *Environmental Science and Engineering* 1 (2019): 305–317.
[2] Syahir, A. Z., N. W. M. Zulkifli, H. H. Masjuki, M. A. Kalam, A. Alabdulkarem, M. Gulzar, L. S. Khuong, and M. H. Harith. "A review on bio-based lubricants and their applications." *Journal of Cleaner Production* 168 (2017): 997–1016.
[3] Srivyas, P. D., and M. S. Charoo. "A review on tribological characterization of lubricants with nano additives for automotive applications." *Tribology in Industry* 40, no. 4 (2018): 594–623.
[4] Quinchia, L. A., M. A. Delgado, T. Reddyhoff, C. Gallegos, and H. A. Spikes. "Tribological studies of potential vegetable oil-based lubricants containing environmentally friendly viscosity modifiers." *Tribology International* 69 (2014): 110–117.
[5] Talib, N., R. M. Nasir, and E. A. Rahim. "Tribological behaviour of modified jatropha oil by mixing hexagonal boron nitride nanoparticles as a bio-based lubricant for machining processes." *Journal of Cleaner Production* 147 (2017): 360–378.
[6] Zulhanafi, P., S. Syahrullail, and M. A. Ahmad. "The tribological performance of hydrodynamic journal bearing using bio-based lubricant." *Tribology in Industry* 42, no. 2 (2020): 278.
[7] Ahmad, M. Ali, I. H. Abdul Razak, N. N. N. A. Fuad, and K. S. Shahrudin. "The effects of different nanoparticles in palm oil olein as solid additives." *Proceedings on Engineering Sciences* 1, no. 1 (2019): 24–30.
[8] Nagendramma, Ponnekanti, and Prashant Kumar. "Eco-friendly multipurpose lubricating greases from vegetable residual oils." *Lubricants* 3, no. 4 (2015): 628–636.
[9] Ortansa, Florea, Marcel Luca, and Steliean Camelia. "Ecological lubricating greases." *Tribology in Industry* 26, no. 1&2 (2004): 52.
[10] Adhvaryu, Atanu, Sevim Z. Erhan, and Joseph M. Perez. "Preparation of soybean oil-based greases: Effect of composition and structure on physical properties." *Journal of Agricultural and Food Chemistry* 52, no. 21 (2004): 6456–6459.

[11] Panchal, Tirth, Dhiraj Chauhan, Merlin Thomas, and Jigar Patel. "Bio based grease A value added product from renewable resources." *Industrial Crops and Products* 63 (2015): 48–52.
[12] Padgurskas, Juozas, Raimundas Rukuiža, Arturas Kupcinskas, and Raimondas Kreivaitis. "Lubrication properties of modified lard and rapeseed oil greases with sodium and lithium thickeners." *Industrial Lubrication and Tribology* 67, no. 6 (2015): 557–563.
[13] Khalil, Waleed, Alaa Mohamed, Mohamed Bayoumi, and T. A. Osman. "Tribological properties of dispersed carbon nanotubes in lubricant." *Fullerenes, Nanotubes, and Carbon Nanostructures* 24, no. 7 (2016): 479–485.
[14] Zhou, Jingfang, Zhishen Wu, Zhijun Zhang, Weimin Liu, and Hongxin Dang. "Study on an antiwear and extreme pressure additive of surface coated LaF3 nanoparticles in liquid paraffin." *Wear* 249, no. 5–6 (2001): 333–337.
[15] Qiu, Sunqing, Zhongrong Zhou, Junxiu Dong, and Guoxu Chen. "Preparation of Ni nanoparticles and evaluation of their tribological performance as potential additives in oils." *Journal of Tribology* 123, no. 3 (2001): 441–443.
[16] Chen, Y. K., O. P. Modi, A. S. Mhay, A. Chrysanthou, and J. M. O'Sullivan. "The effect of different metallic counterface materials and different surface treatments on the wear and friction of polyamide 66 and its composite in rolling–sliding contact." *Wear* 255, no. 1–6 (2003): 714–721.
[17] Rapoport, L., V. Leshchinsky, M. Lvovsky, I. Lapsker, Y. Volovik, Y. Feldman, R. Popovitz-Biro, and R. Tenne. "Superior tribological properties of powder materials with solid lubricant nanoparticles." *Wear* 255, no. 7–12 (2003): 794–800.
[18] Rapoport, Lev, Oleg Nepomnyashchy, Igor Lapsker, Armen Verdyan, Alexey Moshkovich, Yishay Feldman, and Reshef Tenne. "Behavior of fullerene-like WS2 nanoparticles under severe contact conditions." *Wear* 259, no. 1–6 (2005): 703–707.
[19] Rapoport, L., V. Leshchinsky, M. Lvovsky, O. Nepomnyashchy, Y. Volovik, and R. Tenne. "Friction and wear of powdered composites impregnated with WS2 inorganic fullerene-like nanoparticles." *Wear* 252, no. 5–6 (2002): 518–527.
[20] Mohamed, A., T. A. Osman, A. Khattab, and M. Zaki. "Tribological behavior of carbon nanotubes as an additive on lithium grease." *Journal of Tribology* 137, no. 1 (2015): 011801.
[21] Li, Ruiyun, Xing Yang, Deliang Hou, Yongfu Wang, and Junyan Zhang. "Superlubricity of carbon nanostructural films enhanced by graphene nanoscrolls." *Materials Letters* 271 (2020): 127748.
[22] Eswaraiah, Varrla, Venkataraman Sankaranarayanan, and Sundara Ramaprabhu. "Graphene-based engine oil nanofluids for tribological applications." *ACS Applied Materials & Interfaces* 3, no. 11 (2011): 4221–4227.
[23] Dai, Wei, Bassem Kheireddin, Hong Gao, and Hong Liang. "Roles of nanoparticles in oil lubrication." *Tribology International* 102 (2016): 88–98.
[24] Awang, N. W., D. Ramasamy, K. Kadirgama, G. Najafi, and N. A. C. Sidik. "Study on friction and wear of Cellulose Nanocrystal (CNC) nanoparticle as lubricating additive in engine oil." *International Journal of Heat and Mass Transfer* 131 (2019): 1196–1204.
[25] Zhao, Jun, Yingru Li, Junyuan Mao, Yongyong He, and Jianbin Luo. "Synthesis of thermally reduced graphite oxide in sulfuric acid and its application as an efficient lubrication additive." *Tribology International* 116 (2017): 303–309.

[26] Pham, Sang T., Shanhong Wan, Kiet A. Tieu, Ming Ma, Hongtao Zhu, Huynh H. Nguyen, David R. G. Mitchell, and Mitchell J. Nancarrow. "Unusual competitive and synergistic effects of graphite nanoplates in engine oil on the tribofilm formation." *Advanced Materials Interfaces* 6, no. 19 (2019): 1901081.
[27] Omrani, Emad, Pradeep L. Menezes, and Pradeep K. Rohatgi. "Effect of micro- and nano-sized carbonous solid lubricants as oil additives in nanofluid on tribological properties." *Lubricants* 7, no. 3 (2019): 25.
[28] La, Duong Duc, Tuan Ngoc Truong, Thuan Q. Pham, Hoang Tung Vo, Nam The Tran, Tuan Anh Nguyen, Ashok Kumar Nadda et al. "Scalable fabrication of modified graphene nanoplatelets as an effective additive for engine lubricant oil." *Nanomaterials* 10, no. 5 (2020): 877.
[29] Suresha, B., G. Hemanth, A. Rakesh, and K. M. Adarsh. "Tribologicalbehaviour of neem oil with and without graphene nanoplatelets using four-ball tester." *Advances in Tribology* 2020 (2020): 1–11.
[30] Jayadas, N. H., and K. Prabhakaran Nair. "Coconut oil as base oil for industrial lubricants—Evaluation and modification of thermal, oxidative and low temperature properties." *Tribology International* 39, no. 9 (2006): 873–878.
[31] Jayadas, N. H., K. Prabhakaran Nair, and G. Ajithkumar. "Tribological evaluation of coconut oil as an environment-friendly lubricant." *Tribology International* 40, no. 2 (2007): 350–354.

Chapter 5

Nanoadditives-enriched natural-oils-based cutting fluids under minimum quantity lubrication

Wani Khalid Shafi, Sanjay Kumar, M. S. Charoo, and M. Hanief

ABBREVIATION

MQL	Minimum quantity lubrication
NF-MQL	Nanofluid-assisted minimum quantity lubrication
D.O.C	Depth of cut
CNT	Carbon nanotubes
Al_2O_3	Aluminum oxide
MoS_2	Molybdenum disulfide
MWCNTS	Multi-walled carbon nanotubes
C.O.F	Coefficient of friction
CuO	Copper oxide
SiO_2	Silicon dioxide
TiO_2	Titanium dioxide
SiC	Silicon carbide
R_a	Surface roughness
F_t	Tangential force
F_n	Normal force
CHTC	Convective heat transfer coefficient

5.1 INTRODUCTION

Sustainable manufacturing is a major concern in the 21st century. The use of conventional cutting fluids for lubrication and cooling (flood lubrication) of cutting tools leads to ecological degradation, exhaustion of natural resources, and health hazards to the operators. Gupta et al. (2021) reported that the coolants and their management contribute to around 7–17% of the total manufacturing costs, whereas Sreejith and Ngoi (2000) estimated the lubricant costs to be 16–20% of the total product cost. Said et al. (2019) reported that the global utilization of mineral oil-based cutting lubricants in 2016 was around 13,726 million tons, with an annual increment of 1% and presents a serious concern for the preservation of the natural resources. In addition, Talon et al. (2020) reported that the contact of the operators

DOI: 10.1201/9781003717362-5

with the cutting fluids is responsible for 80% of their occupational diseases. Consequently, researchers have devised various alternatives to the problem of conventional machining operations, such as dry machining, cryogenic machining, minimum quantity lubrication (MQL), and solid lubrication, as depicted in Figure 5.1 (Gupta et al., 2021). Minimum quantity lubrication has gained a lot of interest in recent years. It includes a mixture of small oil and compressed air, forming a fine mist or spray that is supplied to the cutting region. Consequently, a very small amount of oil (0.01–2 l/hr) is used in comparison to flood lubrication (500–1000 l/hr). This results in the minimization of resources (less oil exhaustion), reduction in soil and water pollution, and other health benefits by decreasing the contact of operators with the toxic fluids. Moreover, the chips do not get mixed with the oil, thereby increasing recyclability. Different types of MQL are depicted in Figure 5.2. The machining based on MQL reported in different studies is depicted in Table 5.1. Researchers added that petroleum-based cutting oils can be replaced with biodegradable natural oils. Vegetable oils

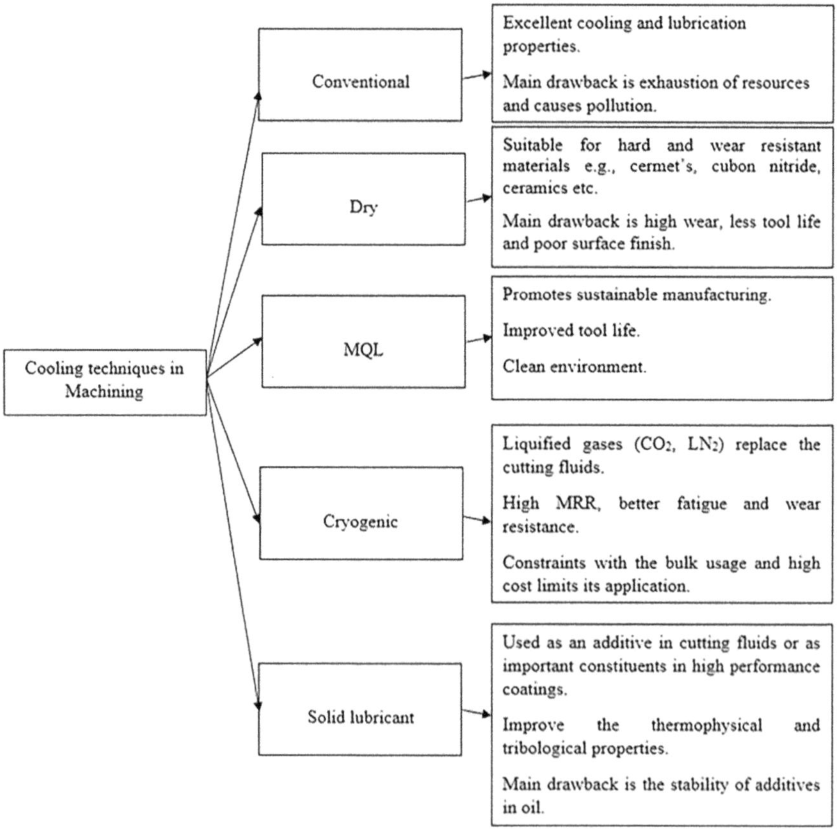

Figure 5.1 Various cooling techniques in machining operations.

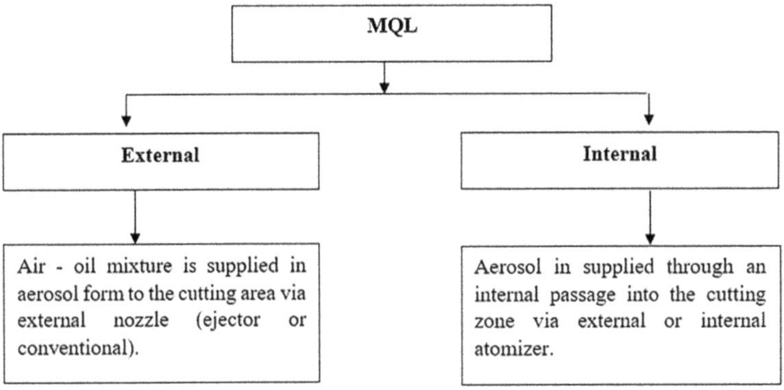

Figure 5.2 Different types of MQL.

possess excellent lubricant characteristics, such as high fire points, high flashpoints, high viscosity indices, and good tribological properties (Shafi et al., 2018). Consequently, this adds to the sustainable manufacturing solutions. Vegetable oils consist of polar and nonpolar groups. The polar groups attach themselves to the metal surfaces, whereas the nonpolar groups prevent the surface–surface interactions by forming monolayer or multimolecular layers between them. The structure of the fatty acids also controls the effectiveness of the protective films that are formed on the surfaces (Shafi et al., 2021). Saturated fatty acids are aligned in a linear manner and result in stronger protective films. The double bonds in unsaturated fatty acids cause a bend in the fatty acid chain, consequently resulting in weaker protective films. Furthermore, the number of double bonds in a fatty chain governs the parameters like viscosity. Unsaturated fatty acids are associated with low viscosity, and saturated fatty acids are associated with high viscosity due to their high melting points (Shafi et al., 2021). Zhang et al. (2016b) reported in their study that high viscosity and high surface tension fluids result in better lubrication, whereas Prabhu et al. (2009) reported that small contact angles can be very advantageous in narrow capillary flows, as in the case of machining. The free fatty acids and the peroxide values control the flashpoints of the vegetable oils, whereas low-temperature properties, like the pour point, are again controlled by the amount of saturated fatty acids present in the oil (Shafi et al., 2021).

Further research has suggested that the introduction of nanoadditives in the cutting fluids results in the improvement of the thermophysical and tribological properties of the fluids (Shafi and Charoo, 2021a, 2021b). Nanoadditives are added to the cutting fluid by a one-step method or a two-step method, and the colloidal fluid is then used as the coolant for machining purposes. Different types of nanoadditives are added to the base oil, viz., metals and their oxides, carbides, sulfides, carbon-based, and rare earth metals. Different nanoadditives have different effects on the base oil

Table 5.1 Reports on minimum quantity lubrication (Dhar et al., 2006; Liao and Lin, 2007; Khan et al., 2009; Tawakoli et al., 2009; Li and Chou, 2010; Sadeghi et al., 2010; Dureja et al., 2015; Rahim et al., 2015; Çakır et al., 2016)

Type of oil	*Type of machining*	*Work–tool material*	*Cutting parameters*	*Findings*
Soluble oil	Turning	AISI – 4340/carbide insert	Cutting velocity = 110 m/min, D.O.C = 1.5 mm, feed rate = 0.16 mm/rev.	Principal flank wear decreased from 500 μm with MQL to 350 μm with wet lubrication, whereas the surface roughness decreased from 6 to 4.25 μm with MQL.
Synthetic ester	End milling	NAK80 hardened steel/carbide insert	Cutting velocity = 300–500 m/min, axial D.O.C = 0.3 mm, radial D.O.C = 5 mm, feed rate = 0.10–0.20 mm/tooth.	MQL outperformed dry machining at all speeds due to oxide film formation.
Vegetable oil	Turning	AISI 9310/uncoated carbide insert	Cutting velocity = 223–483 m/min, mm/rev, D.O.C = 1 mm, feed rate = 0.10–0.18.	MQL machining decreased the average temperature at the chip-tool interface by 10% in comparison to wet lubrication.
Vegetable oil, synthetic oil, Behran cutting oil 34 and 35	Grinding	AISI 4140/grinding wheel	Work speed = 10–40 m/min, D.O.C = 0.005–0.015 mm, wheel speed = 30 m/s,	MQL decreased the cutting forces in comparison to the flood cooling with synthetic oil, depicting the maximum improvement among all the oils. Surface roughness was higher than that of food cooling.
Bluebe	Micro-milling	SKD 61 steels/ uncoated carbide insert	Feed rate = 1–2 μm, D.O.C= 300 μm.	MQL improved the tool flank wear by 60% in comparison to dry milling. Surface roughness remained nearly constant (0.2 μm) at all cutting lengths.

Trimofin-55	Turning	AISI 202/coated carbide insert		Taguchi-based optimization gave a minimum flank wear of 36.63 μm with a surface roughness of 0.74 μm.
LB 8000	Grinding	Hardened 100 Cr6, soft steel (42CrMo4)/Al_2O_3 wheel	Work speed = 2.5–10 m/min D.O.C = 5 μm, wheel speed = 20–30 m/s.	MQL improved the surface roughness of 100 Cr6, whereas 42CrMo4 depicted higher surface roughness in comparison to flood lubrication. MQL decreased the cutting forces, C.O.F, and the specific energy for both the workpieces.
Synthetic ester	Orthogonal cutting	AISI 1045/uncoated carbide insert	Cutting velocity = 350–550 m/min, D.O.C= 2 mm, feed rate = 0.08–0.12 mm/rev.	MQL resulted in the reduction of cutting temperature by 10–30%, whereas cutting forces decreased by 5–28% in comparison to dry conditions.
LubriFluid-F100	Turning	AA7025, AA2024	Cutting velocity = 150–300 m/min, D.O.C = 2.5 mm, feed rate = 0.1–0.2 mm/rev.	The feed rate had the maximum effect on the surface roughness in comparison to other parameters, whereas the material type had the minimum effect on surface roughness.

and depend on various factors, such as shape, concentration, size, chemical structure, and chemical properties of the base oil (Shafi and Charoo, 2018). Nanoadditives increase the heat capacity and result in higher cooling effects. It decreases the friction and wear of work–tool surfaces by (1) formation of a protective layer, (2) polishing action, (3) depositing in the surface asperities, and (4) ball bearing action (Ashraf et al., 2021). The introduction of nanoadditives increases the viscosity of the base oil. Wang et al. (2016) reported that nanofluids with higher viscosity depict better lubricating properties. The low flowability of fluids ensures that the grinding or cutting fluid remains at the contact area for longer durations, consequently decreasing friction and rapid wear of the abrasive particles. However, this may affect the cooling properties of the MQL as poor liquidity dissipates heat in a poor manner. The effectiveness of nanoadditives in the minimum quantity lubrication systems is depicted in Table 5.2.

5.2 NANOFLUID-ASSISTED MINIMUM QUANTITY LUBRICATION-BASED GRINDING

Grinding is a machining process carried out to achieve high accuracy and precision of manufactured parts. The hard grains on the grinding wheel remove a thin layer from the work surface by abrasive action. The specific energy of the grinding process is high, consequently resulting in a higher temperature at the work–tool interface. The increase in temperature is further amplified by the high friction resulting from the relative motion at the work–wheel interface. The temperature at the work–wheel interface can go higher than 1000°C. Higher temperatures result in different internal and external defects, such as surface cracks, surface burn, surface tempering, residual stresses, change in metallographic structure, etc. Another disadvantage of the higher temperatures is the decrease in the hardness of the abrasive grains, which will result in diffusive and bonding wear between the workpiece and the grinding wheel. Consequently, the generated heat must be dissipated quickly from the contact zone.

Li et al. (2017) investigated the grinding force and grinding temperature of a palm oil-based nano-cutting fluid under MQL conditions. CNT nanoparticles were added in eight-volume percentages (0.5–4%). It was observed that palm oil with 2 vol.% CNT concentration depicts the lowest grinding force, grinding temperature, and proportionality coefficient equal to 21.93 N, 109.8 °C, and 42.7%, respectively. The authors reported an increase in the thermal conductivity and viscosity with an increase in the particle concentration. Nanofluids with higher thermal conductivity mean better heat transfer, whereas higher viscosity is suitable for lubrication. However, higher agglomeration of the nanoparticles resulted in the damage of the lubricating film and the obstruction of thermal motion, consequently resulting in poorer properties at high concentrations. Singh et al. (2020a) made a comparative

Table 5.2. Effect of nanoparticles in minimum quantity lubrication (Lee et al., 2010; Setti et al., 2015; Zhang et al., 2016a; Ganesan et al., 2018; Hegab et al., 2018b; Kumar et al., 2017; Dixita et al., 2017; Singh et al., 2018; Akhtar et al., 2020)

Type of oil/nanoparticles	*Type of machining*	*Work–tool material*	*Cutting parameters*	*Findings*
Dialube-30, Dialube-150/Diamond	Grinding	SK-41C/ vitrified CBN wheel	Grinding speed = 800 rpm, D.O.C = 5µm, flow rate = 7.5 cc/hr, feed rate = 120 mm/min.	Normal and tangential forces decreased by 33.2% and 30.3% in comparison to air grinding. Smaller-sized nano diamond particles decreased the surface roughness by 64% and resulted in smoother work surfaces in comparison to air grinding.
Water/graphite, MoS_2, and WS_2	Grinding	Silicon nitride ceramics/ diamond wheel	Wheel speed = 10–20 m/s, MQL flow rate = 150 ml/hr, D.O.C = 10–30 µm.	All the nanofluids depicted lower C.O.F in comparison to dry grinding with WS_2, exhibiting a minimum C.O.F of 0.1980, whereas MoS_2-based nanofluids depicted minimum surface roughness.
Synthetic lipids/MoS_2, MWCNT's, and MoS_2-MWCNT's	Grinding	GH4169-Ni alloy	Wheel speed = 30 m/s, D.O.C= 10 µm, feed rate = 3000 mm/min, MQL flow rate = 50 ml/hr.	Hybrid nanoparticles depicted a better grinding ratio and surface roughness in comparison to individual counterparts and are equal to 0.274 and 0.328 µm, respectively.
Water/Al_2O_3, CuO	Grinding	Ti-6AL-4V	Grinding speed=17 m/s, D.O.C = 5 µm, table speed = 4 m/min.	C.O.F decreased by 32.3% and 42.2% with Al_2O_3 and CuO nanoparticles as compared to dry and wet lubrication. Wet lubrication depicted lower surface roughness among all samples.
Emulsion/MWCNT's	Turning	AISI/uncoated cemented carbide inserts	Cutting speed = 96.2 m/min, feed rate = 0.16 mm/rev, D.O.C = 1 mm.	A decrease in the cutting force, thrust force, and feed forces was observed with NF-MQL reduction in the tool flank wear, and cutting temperatures were also observed.

(Continued)

Table 5.2 (Continued)

Type of oil/nanoparticles	*Type of machining*	*Work–tool material*	*Cutting parameters*	*Findings*
Soybean oil/MWCNT's	Turning	Inconel 625	Feed rate = 0.05–0.15 mm/rev, cutting speed = 15–50 mm/min, D.O.C = 0.3–0.47 mm.	Tool wear under NF-MQL improved by 18.17% and 4.5% in comparison to dry and flood lubrication. The surface finish of the workpiece improved by 55–58% and 4.48% in comparison to dry and wet conditions.
Ethylene glycol/MWCNT's, TiO_2, and Al_2O_3	Turning	AISI 1018/ Tungsten carbide	Feed rate = 0.3 -0.5 mm/rev, cutting speed = 350, 550, and 750 m/min, D.O.C= 0.5–1.5 mm.	TiO_2-based nanofluids depicted maximum thermal conductivity of 0.315 W/m °C at 0.05 wt.% and 50 °C. The lowest cutting temperature was observed at 0.3 wt.% Al_2O_3-based nanofluids. A higher concentration of TiO_2 and Al_2O_3 resulted in the maximum improvement in surface roughness equal to 83% and 29.3%, respectively, in comparison to dry machining.
Vegetable oil/MWCNT's	Turning	Ti-6Al-4V/ coated carbide inserts	Cutting speed = 120–220 m/min, feed rate = 0.1–0.2 mm/rev.	The flank wear and power consumption decreased by 45% and 11.5% at 2 wt.%, respectively, in comparison to pure MQL.
Ethylene glycol/Cu	Turning	H-11 steel/ carbide insert	Cutting speed = 90–109 m/min, feed rate = 0.1–0.2 mm/rev.	The tool wear and surface roughness decreased by 66% and 40% with NF-MQL in comparison to pure MQL. The addition of Cu nanoparticles resulted in the minimization of notch-shaped chips.

investigation of graphene-based nanofluid for Ti-6Al-4V-ELI under different conditions of flood cooling, dry grinding, and MQL. The fluids used for cooling purposes were olive, canola, and sunflower oil. Grinding was carried out at a wheel speed of 22 m/s with a work speed of 3000 mm/min, and the measured results are depicted in Figures 5.3 and 5.4. It is observed from the figures that the surface roughness, tangential force, C.O.F, and grinding energy decrease with the increase in graphene concentration up to an optimal concentration of 1.5 wt.%, beyond which the machining performance started decreasing. It was observed that canola oil +1.5 wt.% graphene depicts the maximum improvement in C.O.F, grinding force, and surface roughness equal to 15.1%, 22.1%, 33.83%, and 16.9%, respectively. The improvement was attributed to the synergistic effect of the graphene nanoparticles and bio-oil that develops a superior lubrication film on the surface.

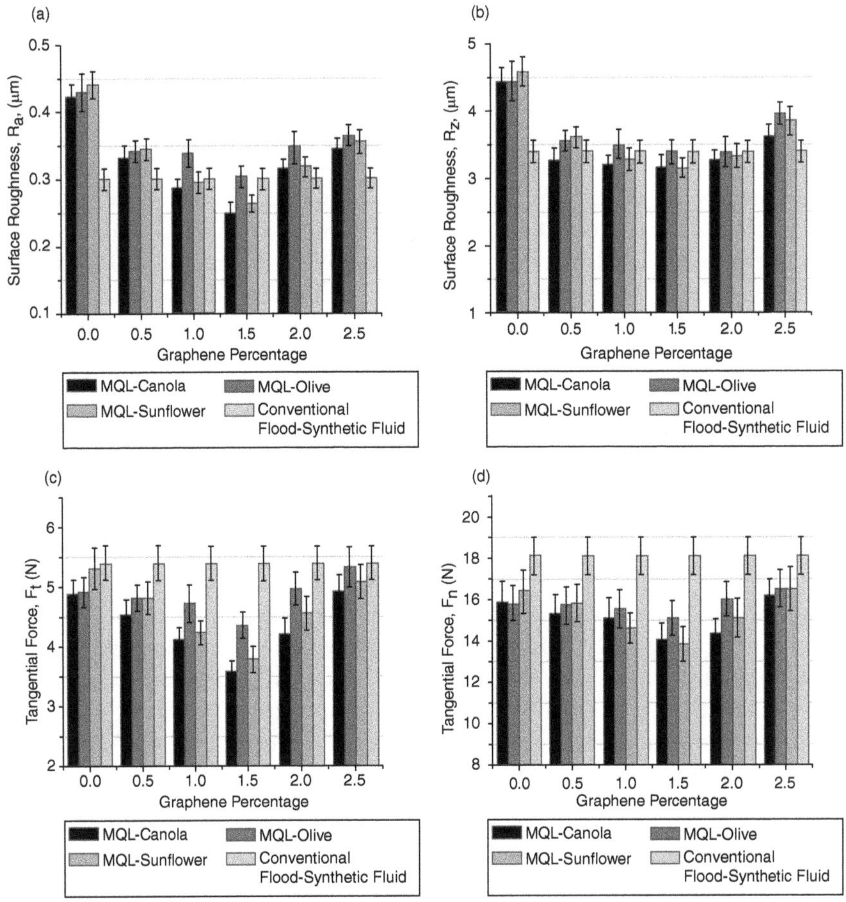

Figure 5.3 Variation of R_a (a), R_z (b), F_t (c), F_n (d), with different weight concentration of graphene in the base oils.

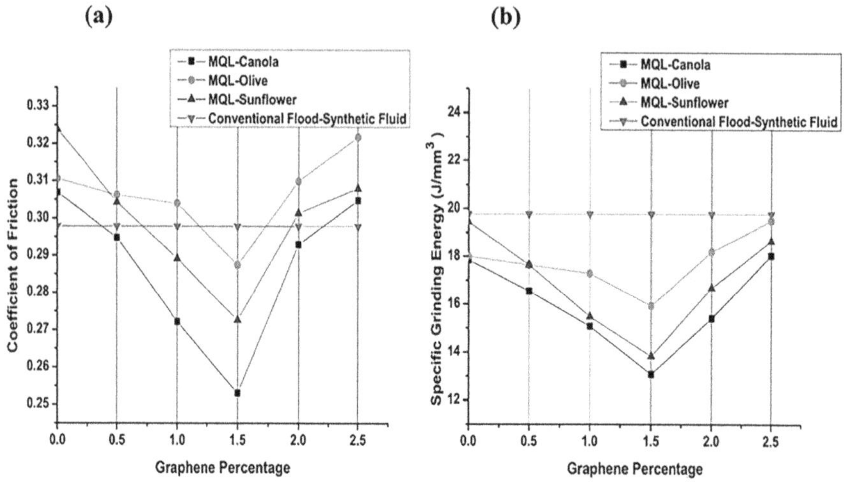

Figure 5.4 Variation of COF (a) and specific grinding energy (b); with different weight concentrations of graphene in the oils.

Shabgard et al. (2017) studied the effectiveness of canola oil – CuO nanofluid under MQL conditions. The authors evaluated different grinding characteristics, such as grinding force, surface roughness, and surface temperature. It was observed that deformation and sliding were the main reasons for the increase in C.O.F under pure MQL conditions. The introduction of nanoparticles in the oil resulted in a decrease in the grinding ratio (G-ratio). This was attributed to the tribofilm formation of the CuO nanoparticles on the wheel–work, thereby depicting higher lubrication capabilities. Furthermore, CHTC with the canola – CuO nanofluids also depicted lower grinding temperatures in comparison to the pure MQL. The rolling action of the nanofluids resulted in the lowering of sliding forces and generated less heat during the grinding process. Furthermore, the lubricating film on the surface also reduced the energy partition of the workpiece and depicted lower grinding temperatures. Virdi et al. (2021) also mixed CuO nanoparticles in vegetable oil (palm oil) and studied different parameters, such as grinding forces and specific grinding energy, in different types of grinding. The authors reported that the addition of 0.5 wt.% CuO in palm under MQL conditions depicted a better surface finish than the pure oil, owing to the rolling and polishing action of nanoparticles on the work surface. Furthermore, 1 wt.% concentration resulted in a higher MRR with lesser energy consumption in comparison to flood and dry grinding.

Wang et al. (2016) studied the effect of six different nanoparticles (MoS_2, CNT's, SiO_2, diamond, ZrO_2 and Al_2O_3) on the lubrication performance of GH4169 Ni-based alloy. It was observed that all the NF-MQL-based nanofluids depict less C.O.F and specific grinding energy in comparison to flood lubrication. Al_2O_3-based nanofluids depicted a minimum C.O.F equal to

0.348 in comparison to 0.583 with flood lubrication. Furthermore, Al_2O_3-based NF-MQL depicted minimum specific cutting energy corresponding to 35.7% improvement in comparison to flood lubrication. The authors reported that nanoparticles get adhered to the wheel-workpiece by penetration or adsorption and form a low shear strength oil film. Furthermore, the nanoparticles penetrate better into the contact area under high pressure and reduce the frictional force and specific grinding energy. The rolling and mending mechanism further improved the lubrication effect. Zhang et al. (2019) mixed the hybrid nanoparticles (Al_2O_3/SiC) in different types of synthetic and vegetable-based oils (castor, palm) and studied their effect on grinding force and surface roughness under MQL conditions. The nanoparticles were 70 nm in size, and the workpiece was made up of GH4169Ni-based alloy. It was observed that the synthetic oil-based nanofluid depicts a minimum microscopic frictional coefficient, followed by castor oil + Al_2O_3/SiC particles equal to 0.173 and 0.218, respectively, at 1 vol.%.

Zhang et al. (2015a) made a comparative investigation of frictional and wear properties under different machining conditions (dry, flood, MQL, and NF-MQL). Minimum specific grinding energy was observed with flood grinding equal to 29.8 J/mm^3 in comparison to 45.5 J/mm^3 under MQL. The introduction of MoS_2 under MQL conditions further decreased the specific energy to 32.7 J/mm^3 and comes very close to the flood grinding. Similar trends were observed with C.O.F, where the minimum C.O.F was depicted under flood grinding, followed by NF-MQL. Pal et al. (2020b) mixed MoS_2 in the rapeseed oil and studied its characteristics, such as grinding force, surface roughness, and temperature by varying flow rate and pressure. The authors reported that the NF-MQL-based coolant improved the specific normal and tangential forces by 43% and 33%, respectively, at 1 wt.%. The improvement was attributed to the rolling and film formation capability of MoS_2 nanoparticles on the surface. This also resulted in less heat generation at the contact interface. Furthermore, surface analysis of the workpiece revealed the minimization of microcracks, deep furrows, and adhesion as observed with pure MQL.

Jia et al. (2014) measured the frictional coefficient, specific cutting energy, and grinding force under dry, flood, MQL, and NF-MQL conditions for a soybean-MoS_2 coolant. The authors reported a decrease of 19% in tangential force under NF-MQL in comparison to the MQL. Furthermore, the specific cutting energy and C.O.F also decreased by 30.4% and 20.3%, respectively, in comparison to MQL. MoS_2 nanoparticles also improved the heat exchange at the grinding zone, thereby leading to better cooling effects. Ibrahim et al. investigated the effectiveness of bio-oil (palm oil) mixed with graphene nanoplatelets (thickness: 3–10 nm) under MQL conditions for grinding. The authors reported a decrease of 74.5% in specific grinding energy at 0.1 wt.% concentration in comparison to pure palm oil. The frictional coefficient decreased by 75.2%, whereas the wear rate decreased by 71.9% at a similar concentration. The authors reported that the introduction of nanoparticles increases the contact angle and consequently decreases the wettability of nanofluids. As a result, direct contact between the wheel and worksurface is present in some regions, which leads to an increase in the surface roughness. However, this effect was also diminished by the nanoparticles by forming an antifriction layer between the wheel and the worksurface.

Zhang et al. (2015b) introduced MoS_2 nanoparticles in soybean oil, palm oil, and rapeseed oil and studied their grinding characteristics with 45 steel workpieces under MQL conditions. The authors reported that minimum C.O.F is depicted by palm oil +2% MoS_2 equal to 0.3 and comes close to the flood lubrication, where C.O.F is equal to 0.28. In addition, the minimum specific cutting energy was observed at a similar concentration equal to 65.7 J/mm^3 in comparison to 76.1 J/mm^3 with pure palm oil.

5.3 NANOFLUID-ASSISTED MINIMUM QUANTITY LUBRICATION-BASED MILLING

The use of the MQL technique in milling operations is considered a viable option in the machining of materials that are difficult to cut. The temperature in the work area increases greatly, and the introduction of nanoparticles in the MQL fluid can result in higher cooling effects. Sen et al. (2020) investigated the effect of alumina–palm oil-based nanofluids on the frictional and wear properties of Inconel 690 during the end milling machining process and compared its results with the pure MQL and flooded milling conditions. The authors reported a decrease of 19.35% in tool wear with NF-MQL in comparison to the flooded milling. Al_2O_3 nanoparticles established a protective film and resulted in a smoother work surface. Sugiantoro (2018) also studied the effect of Al_2O_3, SiO_2, and CuO on the milling characteristics under MQL and NF-MQL conditions. Al_2O_3 nanoparticles were mixed with coconut oil, whereas SiO_2 nanoparticles were mixed with palm oil. It was observed that palm oil mixed with SiO_2 nanoparticles depicts a minimum temperature at the cutting area, minimum surface roughness, and power consumption in comparison to other nanoparticles.

Yin et al. (2018) introduced six different nanoparticles in cottonseed oil and studied the milling characteristics of Ti-6AL-AV under MQL and NF-MQL conditions. All the NF-MQL-based cutting fluids depicted lesser milling forces, C.O.F., and surface roughness in comparison to the base oil. Al_2O_3 depicted the maximum improvement in cutting forces (F_x and F_y) of 10.34% and 52.48%, respectively, and graphite NF-MQL with a minimum improvement of 5.17% and 11.39%, respectively. Al_2O_3-based NF-MQL also depicted maximum improvement in C.O.F equal to 47.7% as compared to pure cottonseed oil, followed by SiO_2 NF-MQL equal to 46.1%. The authors concluded that Al_2O_3 nanoparticles formed a protective film by getting deposited on the surface and thereby reducing the shear stress, consequently reducing the friction and wear. Bai et al. (2021) also studied the cutting forces of cottonseed oil mixed with Al_2O_3 nanoparticles and reported that the Al_2O_3 decreased the F_x and F_y cutting forces by 29.5% and 58.85%, respectively, in comparison to pure cottonseed oil. Further, the minimum surface roughness was also depicted by Al_2O_3-based nano-cutting fluids with an improvement of 66.8% in comparison to pure oil. The improvement was again attributed to the protective film formed by the nanoparticles on the work surface. The authors also reported that Al_2O_3 and SiO_2-based nanofluids were more effective owing to their higher viscosity and concluded that higher viscosity nanofluids result in a better lubricating effect.

Gunan et al. (2020) investigated the tool life of Hastelloy (C276) under MQL and NF-MQL conditions at varying speeds (60, 75, and 90 m/min) and feed rates (0–0.20 mm/rev). It was observed that the tool life decreased by 23% at 1 vol.% Al_2O_3 concentration. Furthermore, a decrease in the adhesive wear was observed at a concentration of 1 and 1.5 vol.%, indicating the effectiveness of nanoparticles in oil. Qingan et al. (2018) introduced Al_2O_3 nanoparticles in oil and evaluated different milling characteristics (milling force, C.O.F, surface roughness, and specific energy) of 45 steel under dry, flood, MQL, and NF-MQL conditions. NF-MQL depicted minimum milling force, C.O.F, and surface roughness with an improvement of 38.46%, 38.15%, and 20.28% over dry milling. The C.O.F observed with NF-MQL was even better than flood lubrication, with an improvement of 14% over it.

Singh et al. (2019a) studied the effect of carbon-based nanoparticles (MWCNTs) on the milling properties of Inconel 625. MWCNTs decreased the surface roughness and tool wear by 17.33% and 11.48%, respectively, in comparison to pure MQL at a cutting speed, depth of cut, and feed rates of 47 m/min, 0.20 mm, and 0.05 mm/tooth, respectively. Huang et al. (2016b) introduced MWCNTs in vegetable oil and reported that NF-MQL depicted better properties than MQL and led to significant improvement in cutting force, temperature, surface roughness, and tool wear. The cutting temperature, surface roughness, and cutting forces decreased by 10.8%, 8.1%, and 4.1%, respectively.

Kim et al. (2017) investigated the micro-end milling characteristics of Ti-6Al-4V under dry, MQL, and NF-MQL with chilly CO_2. The base fluid was vegetable oil with diamond nanoparticles mixed in it. It was reported that lower concentrations (0.1 wt.%) supplied with chilly CO_2 are most effective in improving the milling characteristics. The average resultant force decreased from 8.1 N under MQL to 5 N with NF-MQL, whereas the tool wear decreased from 21.5 to 13 μm with NF-MQL. Highly concentrated (1 wt.%) nanofluids resulted in better surface quality with an average surface roughness of 0.11 μm in comparison to 0.32 μm under pure MQL conditions.

Dong et al. (2019) also reported the milling characteristics of Ti-6Al-4V under NF-MQL conditions with cottonseed as a base oil and Al_2O_3, SiO_2, and MoS_2 nanoparticles as nanoadditives. The authors reported that Al_2O_3

and SiO_2 nanoparticles are the most effective additives in cottonseed oil. NF-MQL with SiO_2 nanoparticles depicted lower milling forces (43.6 N) and the lowest milling temperature. The minimum surface roughness was observed with Al_2O_3 NF-MQL conditions and was 66.7% lower than pure cottonseed oil. Uysal et al. (2015) examined the milling characteristics of martensitic steels (AISI 420) using vegetable oil mixed with MoS_2 nanoparticles as nano-cutting fluid. The authors reported a decrease of 16.8 and 19.9% in tool wear with NF-MQL in comparison to 9.8% and 15.5% at 20 and 40 ml/hr with pure MQL. Furthermore, the surface improved greatly with NF-MQL equal to 27.5% and 16.7% at 20 and 40 ml/hr in comparison to the pure MQL.

Yuan et al. (2018) made a comparative investigation between two vegetable oils (canola oil and soybean oil) mixed with three different nanoparticles (Cu, SiC, and diamond) on the milling characteristics of 7050 aluminum alloy. It was observed that all the nanoparticles reduce the cutting force and surface roughness of the workpiece, with diamond particles yielding the minimum cutting forces. The C.O.F for soybean oil and canola oil with diamond nanoparticles decreased by 1.2% and 1.9%, respectively, as compared to pure MQL. The minimum surface was obtained with canola oil mixed with SiC equal to 616 µm.

Osman et al. (2020) mixed h-BN nanoparticles (80–100 nm) in fatty acid ester and investigated the slot milling of the Ti-6Al-4V workpiece under NF-MQL conditions. It was observed that NF-MQL resulted in small cutting forces with an improvement of 24.75% in surface roughness. Minh et al. (2017) investigated the soybean oil nano-cutting fluid for hard milling of 60 Si_2Mn steel. The authors reported an improvement of 177% in tool life as compared to pure MQL and attributed it to the rolling effect of Al_2O_3 nanoparticles.

5.4 NANOFLUID-ASSISTED MINIMUM QUANTITY LUBRICATION-BASED DRILLING

Chatha et al. (2016) examined the drilling characteristics of aluminum alloy under different machining lubrication scenarios. Al_2O_3 nanoparticles (20 nm) were mixed in soybean oil, and drilling cutting force, surface roughness, and tool wear were measured. The authors reported that the NF-MQL greatly reduced the cutting forces owing to the rolling action of spherical Al_2O_3 nanoparticles at the tool–work interface and consequently resulted in lower frictional forces. NF-MQL also eliminated the chips and burrs and resulted in smoother work surfaces.

Nam et al. (2011) investigated the effect of diamond nanoparticles mixed in bio-oil (ricebran oil) and studied their micro-drilling characteristics of 6061 aluminum alloy under dry, MQL, and NF-MQL conditions. It was reported that the drilling torque and the thrust force were reduced by 68.7%

and 69.1% at 2 vol.% concentration in comparison to the dry machining conditions. The NF-MQL samples were able to drill 150 holes, whereas lubrication with dry air failed at the drilling of the 87th hole. Furthermore, the rolling effect of nanoparticles removed the chips and burns and decreased the adherence of chips on the micro drill.

Mosleh et al. (2019) studied the tribological properties of Ti-6Al-4V using orbital drilling and four-ball testers under NF-MQL conditions. MoS_2 and h-BN nanoparticles are mixed in Boelube, and orbital drilling was carried out at 1 m/s with a lubricant flow rate of 2 cm^3/hr and an air flow rate of 944 cm^3/s. The authors reported that the MoS_2-based nano-cutting fluid resulted in less material transfer and edge build-up of the titanium film on the carbide tool. Four-ball tests revealed a decrease in the frictional coefficient, wear, and surface temperature with both the nanofluids. The tool wear decreased by 43.75% and 41.67% with MoS_2 and h-BN in comparison to the pure oil. The authors concluded that the performance of nanoparticles can further be enhanced by decreasing the size of nanoparticles, which resulted in an increase in surface area for nanoparticle–surface interaction at the interface.

Huang et al. (2016a) also reported the effect of diamond nanoparticles on the micro-drilling characteristics of 7075 – T6 alloy lubricated with bio-oil. The authors reported an improvement in the drilling temperatures at 2 wt.%, equal to 18.3% in comparison to dry milling. The average thrust force and torque at 400 holes decreased by 13.7 and 13.3%, respectively, in comparison to pure MQL. Moreover, the NF-MQL-based fluids resulted in smoother surfaces with fewer burrs around the holes.

Khanafer et al. (2020) reported the micro-drilling of Inconel 718 with rapeseed oil mixed with Al_2O_3 as a cutting fluid. The authors observed a decrease in the thrust forces with NF-MQL in comparison to flood and MQL and reported that the nanoparticles act as a spacer between the tool and workpiece, thereby reducing friction. Furthermore, the protective film formation by nanoparticles also resulted in the lowest wear among the three fluids. SEM analysis of the workpiece under NF-MQL further revealed a smoother surface with fewer abrasive marks and fewer fractured edges.

Pal et al. (2020a) reported an improvement of 27.4, 33.8, and 51.7% in thrust force, surface roughness, and frictional coefficient in comparison to the pure MQL at the drilling of the 30th hole. The authors attributed it to the stable protective film formation by graphene nanoparticles on the interface. Furthermore, the adhesive wear between the tool and work decreased significantly at 1.5 wt.% in comparison to the pure MQL. The authors concluded that the higher viscosity of nanofluids results in the rolling mechanism of nanoparticles on the workpiece, consequently, improving C.O.F and wear.

Pal et al. (2021) also studied the effect of MoS_2 nanoparticles on AISI stainless steel under different machining lubrication conditions. The authors reported the maximum improvement in drilling characteristics at 1.5 wt.%

with 43.2% and 68.4% in thrust force and torque, respectively, whereas the surface roughness and frictional coefficient improved by 56.8% and 41.6%, respectively. MoS_2 gets adsorbed on the interface and provides the lubrication effect. In addition, MoS_2 nanoparticles increased the strength of the oil film and resulted in better cooling and lubrication performance.

5.5 NANOFLUID-ASSISTED MINIMUM QUANTITY LUBRICATION-BASED TURNING

Turning is one of the most common machining operations in manufacturing science. It involves a single-point cutting tool that removes unwanted material in the form of chips from the work surface. The effectiveness of the turning operation depends on the various parameters, such as cutting velocity, depth of cut, and feed rate, that govern the surface quality and wear rate of the workpiece. The NF-MQL has been reported to have effective cooling and lubrication properties that improve the machining properties of the turning operations. Anandan et al. (2020) examined the effect of silver nanoparticle-based cutting fluid under MQL on the turning of titanium under different cutting conditions. It was found that using silver nanoparticle-based cutting fluid greatly reduced the surface roughness and was 60% less than the flood lubrication. The authors also reported a reduction of 34% in cutting temperature with flood lubrication and 18% with oil lubrication owing to the increase in the thermal conductivity of cutting fluids due to the addition of silver nanoparticles. This resulted in excellent heat transfer capability and consequently a reduction in the cutting temperature. Eltaggaz et al. (2018) investigated the effect of (Al_2O_3) gamma nanoparticles on the tool wear behavior during the cutting of austempered ductile iron. It was noted that the MQL nanofluid at the cutting speed of 120 m/min and feed rate of 0.2 mm/rev depicted the best tool life in comparison to the MQL. The authors ascribed it to the tribofilm growth along the tool–workpiece interface zone and nanofluid film that reduced the coefficient of friction. Padmini et al. (2016) examined the machining performance of vegetable oil-based nanofluid during the turning of AISI 1040 steel under MQL. The nanofluid samples were developed by dispersing the varying concentrations of molybdenum disulfide nanoparticles (MoS_2) in coconut oil, sesame oil, and canola oil. It was observed that coconut oil mixed with 0.5 % MoS_2 depicted the maximum improvement among the three oils. The cutting forces, temperatures, tool wear, and surface roughness were reduced by 37%, 21%, 44%, and 39%, as shown in Figure 5.5.

Singh et al. (2018) studied the effect of carbon nanotubes in vegetable oil for minimizing the tool wear and surface roughness under different machining conditions of Inconel 625. The use of NF-MQL improved the surface quality and resulted in uniform flank wear across the tool's cutting edge. It was observed that the tool wear was lowered by 18.17 % and 4.54% and

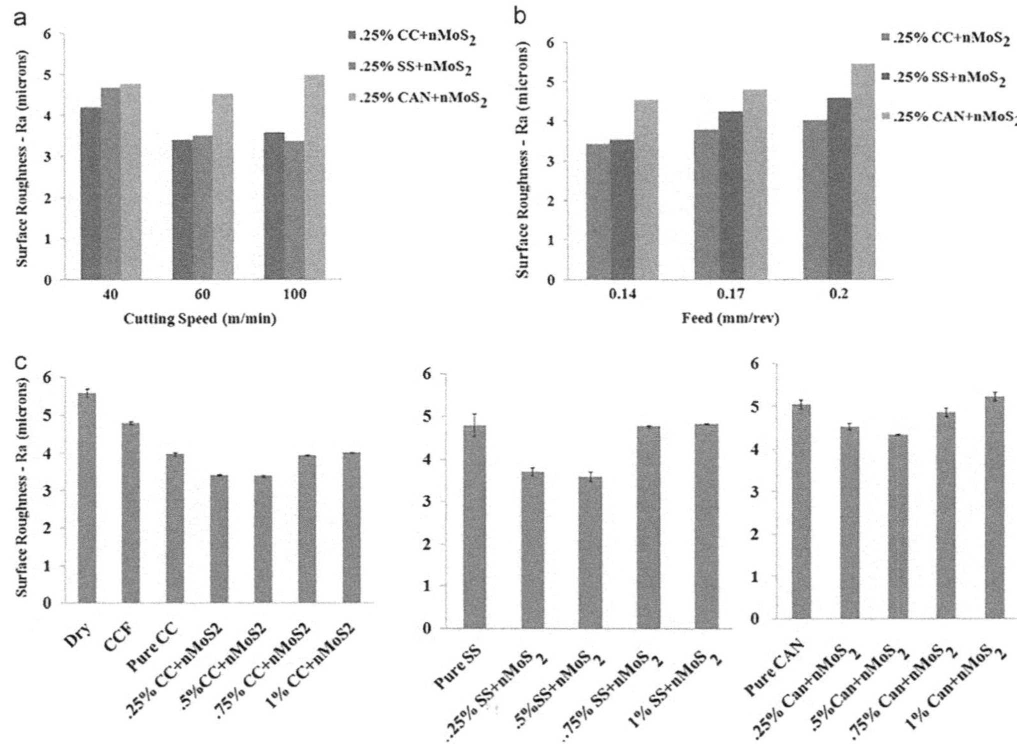

Figure 5.5 (a) Variation of surface roughness with speed (feed: 0.14 mm/rev; time: 16 min; D.O.C: 0.5 mm; npi: 0.25%). (b) Variation of surface roughness with feed (speed: 60 m/min; time: 16 min; D.O.C: 0.5 mm; npi: 0.25%). (c) Variation of surface roughness under varying npi of $nMoS_2$ (0–1%) in corresponding oils (Coconut, Sesame and canola) (speed: 60 m/min; feed: 0.14 mm/rev; D.O.C: 0.5 mm).

the surface finish was enhanced by 55.58 % and 5.48% during the turning of Inconel 625 under NF-MQL as compared to dry machining and wet (flood) cutting conditions. It was reported that abrasion and adhesive wear were the main wear mechanisms in dry cutting, whereas adhesive wear and BUE are prominent under flood coolant conditions. Muhammad et al. (2019) analyzed the effect of cryogenic cooling environment and hybrid nanofluid-based MQL techniques on the turning of Ti–6Al–4V. They synthesized the hybrid nanofluid by adding alumina (Al_2O_3) along with multi-walled carbon nanotubes (MWCNTs) in vegetable oil. In comparison to cryogenic cooling, hybrid nanoadditives decreased the average surface roughness and cutting force by 8.72% and 11.8%, respectively, whereas tool life was enhanced by 23%. It was further observed that the cryogenic approach demonstrated a reduction of 11.2% in cutting temperature in comparison to the MQL-hybrid nanofluids at varied cutting speeds and feed rates. Hegab et al. (2018a) mixed MWCNTs in vegetable oil (ECOLUBRIC) and examined their effect on the turning of Ti-6Al-4V under different cutting design variables. It was observed that using 4 wt.% MWCNTs reduces surface roughness by 38%, whereas 2 wt.% MWCNT significantly improved the surface quality by 50% in comparison to pure MQL. The authors ascribed it to the development of tribofilm along the chip-tool interface zone that resulted in the enhancement in tool wear and surface quality.

Singh et al. (2019c) mixed graphene in vegetable oils and investigated the machining performance of textured carbide tools and titanium (Ti-6Al-4V) alloy under MQL and NF-MQL conditions, and compared their performance under dry and wet conditions. Experimental results demonstrated that NF-MQL provided the longest tool life, the lowest cutting forces, and the lowest cutting temperature. In comparison to dry circumstances, tool life, main cutting force, and cutting temperature improved by 178–190%, 36–40%, and 31–42%, respectively, under NF-MQL at varying cutting speeds. They concluded that the NF-MQL produced the best turning results owing to the increased thermal conductivity and the shearing force provided by graphene on the various contact surfaces of the tool. Singh et al. (2020b) in another study reported the surface roughness, chip morphology, and lubrication mechanisms of the graphene nanoparticles while turning Ti–6Al–4V under dry, MQL, and NF-MQL. They found that graphene-based cutting fluid gives the best performance by increasing the shear angle (11–30%) and reducing the coefficient of friction (16–39%). The Ra under the NF-MQL environment decreased by 41–53% and 58–68% in comparison to the pure MQL and dry turning conditions at various cutting speeds. The lubrication mechanisms are depicted in Figure 5.6. It was observed that the graphene-based MQL gives remarkable performance due to its outstanding thermal conductivity and shearing behavior. The presence of graphene was determined by the EDS analysis, which confirmed the presence of carbon on the cutting edge.

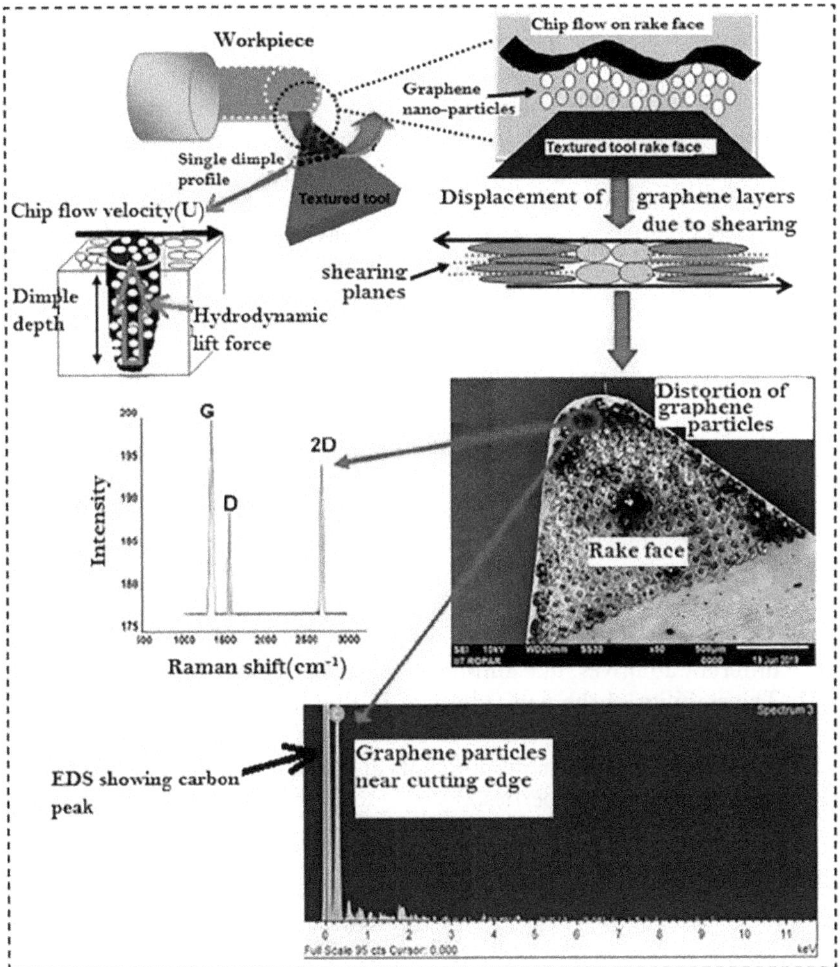

Figure 5.6 Lubrication mechanism behind graphene mixed oil at 180 m/min.

Singh et al. (2019b) also measured the turning performance of AISI 4340 by using pure canola oil and graphene-based canola oil under different cutting conditions. They observed that the texturing on the tools lowered the tool–chip contact length, which helped in keeping the coolant/lubricant on the rake face. It was found that there was a considerable reduction in cutting temperature by using graphene-based canola oil, and it was less than 18–45% in comparison to pure MQL.

Su et al. (2016) examined the influence of nanofluid MQL on cutting force and temperature in turning of AISI 1045 under graphite-based MQL conditions. The nanofluid was prepared by using organic oil and ester oil as base fluids. They observed that NF-MQL considerably lowered the cutting temperature and the cutting force. The maximum enhancement for LB2000 and PriEco6000 oil with graphite nanoparticles at 0.5% was observed at high speeds (96 m/min) and was equal to 26% and 21% in comparison to dry conditions. A similar concentration also resulted in the maximum improvement in the cutting forces. Gong et al. (2020) investigated the machining performance of the Inconel 718 alloy under the MQL based on graphene nanofluids. The nanofluids were formulated using two different sizes of graphene nanoplatelets (5 and 15 μm). Experimental results revealed that the nanofluids with the small-sized enhanced the surface quality by reducing the surface roughness (6%–15%) at the highest and lowest cutting speeds, respectively.

5.6 FUTURE SCOPE

This chapter encourages the use of natural oil-based cutting fluids in machining operations. The sustainability of machining operations can further be improved by employing nanofluid-based minimum quantity lubrication. However, a few challenges need to be overcome for the application of NF-MQL in the industries.

1. The oxidation stability of the bio-oils needs to be improved by adding different additives, like antioxidants, or by chemical modifications.
2. The stability of the nanoadditives in the bio-oils needs to be investigated so that highly stable nanolubricants can be stored for longer durations. Unstable nanolubricants also result in poor thermophysical and tribological properties.

REFERENCES

Akhtar, A. Z., Rahman, M., Kadirgama, K. & Maleque, M. Effect of TiO2 and Al2O3-ethylene glycol-based nanofluids on cutting temperature and surface roughness during turning process of AISI 1018. *IOP Conference Series: Materials Science and Engineering*, 2020. IOP Publishing, 052033. Available at Doi: https://doi.org/10.1088/1757-899X/736/5/052033

Anandan, V., Babu, M. N., Muthukrishnan, N. & Babu, M. D. 2020. Performance of silver nanofluids with minimum quantity lubrication in turning on titanium: a phase to green manufacturing. *Journal of the Brazilian Society of Mechanical Sciences and Engineering*, 42(4), pp 1–15. Available at Doi: https://doi.org/10.1007/s40430-020-02277-7

Ashraf, A., Shafi, W. K., Ul Haq, M. I. & Raina, A. 2021. Dispersion stability of nano additives in lubricating oils–an overview of mechanisms, theories and methodologies. *Tribology - Materials, Surfaces and & Interfaces*, pp 1–23. Available at Doi: https://doi.org/10.1080/17515831.2021.1981720

Bai, X., Jiang, J., Li, C., Dong, L., Duan, Z. & Zhang, H. 2021. Lubrication performance of different concentrations of Al_2O_3 nanofluids on minimum quantity lubrication milling. Available at Doi: https://doi.org/10.21203/rs.3.rs-230176/v1

Çakır, A., Yağmur, S., Kavak, N., Küçüktürk, G. & Şeker, U. 2016. The effect of minimum quantity lubrication under different parameters in the turning of AA7075 and AA2024 aluminium alloys. *The International Journal of Advanced Manufacturing Technology*, 84(9), pp 2515–2521. Available at Doi: https://doi.org/10.1007/s00170-015-7878-4

Chatha, S. S., Pal, A. & Singh, T. 2016. Performance evaluation of aluminium 6063 drilling under the influence of nanofluid minimum quantity lubrication. *Journal of Cleaner Production*, 137(537–545). Available at Doi: https://doi.org/10.1016/j.jclepro.2016.07.139

Dhar, N., Kamruzzaman, M. & Ahmed, M. 2006. Effect of minimum quantity lubrication (MQL) on tool wear and surface roughness in turning AISI-4340 steel. *Journal of Materials Processing Technology*, 172(2), pp 299–304. Available at Doi: https://doi.org/10.1016/j.jmatprotec.2005.09.022

Dixita, R., Mandala, A. & Dasa, A. K. 2017. *Effect of MWCNT Mixed Nanofluid in Turning Operation Using Minimum Quantity Lubrication*. Available at ISBN: 978-93-80689-28-9

Dong, L., Li, C., Xiufang, B., Mingge, Z., Qiang, Q., Qingan, Y., Xiaojie, L. & Li, L. 2019. Analysis of the cooling performance of Ti–6Al–4V in minimum quantity lubricant milling with different nanoparticles. *The International Journal of Advanced Manufacturing Technology*, 103(5–8), pp 2197–2206. Available at Doi: https://doi.org/10.1007/s00170-019-03466-y

Dureja, J., Singh, R., Singh, T., Singh, P., Dogra, M. & Bhatti, M. S. 2015. Performance evaluation of coated carbide tool in machining of stainless steel (AISI 202) under minimum quantity lubrication (MQL). *International Journal of Precision Engineering and Manufacturing-Green Technology*, 2(2), pp 123–129. Available at Doi: https://doi.org/10.1007/s40684-015-0016-9

Eltaggaz, A., Hegab, H., Deiab, I. & Kishawy, H. 2018. Hybrid nano-fluid-minimum quantity lubrication strategy for machining austempered ductile iron (ADI). *International Journal on Interactive Design and Manufacturing (IJIDeM)*, 12(4), pp 1273–1281. Available at Doi: https://doi.org/10.1007/s12008-018-0491-7

Ganesan, K., Babu, M. N., Santhanakumar, M. & Muthukrishnan, N. 2018. Experimental investigation of copper nanofluid based minimum quantity lubrication in turning of H 11 steel. *Journal of the Brazilian Society of Mechanical Sciences and Engineering*, 40(3), pp 1–17. Available at Doi: https://doi.org/10.1007/s40430-018-1093-9

Gong, L., Bertolini, R., Ghiotti, A., He, N. & Bruschi, S. 2020. Sustainable turning of Inconel 718 nickel alloy using MQL strategy based on graphene nanofluids. *The International Journal of Advanced Manufacturing Technology*, 108(9), pp 3159–3174. Available at Doi: https://doi.org/10.1007/s00170-020-05626-x

Günan, F., Kıvak, T., Yıldırım, Ç. V. & Sarıkaya, M. 2020. Performance evaluation of MQL with AL2O3 mixed nanofluids prepared at different concentrations in milling of Hastelloy C276 alloy. *Journal of Materials Research and Technology*, 9(5), pp 10386–10400. Available at Doi: https://doi.org/10.1016/j.jmrt.2020.07.018

Gupta, M. K., Khan, A. M., Song, Q., Liu, Z., Khalid, Q. S., Jamil, M., Kuntoğlu, M., Usca, Ü. A., Sarıkaya, M. & Pimenov, D. Y. 2021. A review on conventional and advanced minimum quantity lubrication approaches on performance measures of grinding process. *The International Journal of Advanced Manufacturing Technology*, 117(3), pp 729–750. Available at Doi: https://doi.org/10.1007/s00170-021-07785-x

Hegab, H., Kishawy, H., Gadallah, M., Umer, U. & Deiab, I. 2018a. On machining of Ti-6Al-4V using multi-walled carbon nanotubes-based nano-fluid under minimum quantity lubrication. *International Journal of Advanced Manufacturing Technology*, 97, 1593–1603. Available at Doi: https://doi.org/10.1007/s00170-018-2028-4

Hegab, H., Umer, U., Deiab, I. & Kishawy, H. 2018b. Performance evaluation of Ti-6Al-4V machining using nano-cutting fluids under minimum quantity lubrication. *International Journal of Advanced Manufacturing Technology*, 95, 9–12. Available at Doi: https://link.springer.com/article/10.1007%2Fs00170-017-1527-z

Huang, W.-T., Wu, D.-H. & Chen, J.-T. 2016a. Robust design of using nanofluid/MQL in micro-drilling. *The International Journal of Advanced Manufacturing Technology*, 85(9), pp 2155–2161. Available at Doi: https://doi.org/10.1007/s00170-015-7382-x

Huang, W.-T., Wu, D.-H., Lin, S.-P. & Liu, W.-S. 2016b. A combined minimum quantity lubrication and MWCNT cutting fluid approach for SKD 11 end milling. *The International Journal of Advanced Manufacturing Technology*, 84(5–8), pp 1697–1704. Available at Doi: https://doi.org/10.1007/s00170-015-7770-2

Jia, D., Li, C., Zhang, D., Zhang, Y. & Zhang, X. 2014. Experimental verification of nanoparticle jet minimum quantity lubrication effectiveness in grinding. *Journal of Nanoparticle Research*, 16(12), pp 1–15. Available at Doi: https://doi.org/10.1007/s11051-014-2758-7

Khan, M., Mithu, M. & Dhar, N. R. 2009. Effects of minimum quantity lubrication on turning AISI 9310 alloy steel using vegetable oil-based cutting fluid. *Journal of Materials Processing Technology*, 209(15–16), pp 5573–5583. Available at Doi: https://doi.org/10.1016/j.jmatprotec.2009.05.014

Khanafer, K., Eltaggaz, A., Deiab, I., Agarwal, H. & Abdul-Latif, A. 2020. Toward sustainable micro-drilling of Inconel 718 superalloy using MQL-Nanofluid. *The International Journal of Advanced Manufacturing Technology*, 107(7), pp 3459–3469. Available at Doi: https://doi.org/10.1007/s00170-020-05112-4

Kim, J. S., Kim, J. W. & Lee, S. W. 2017. Experimental characterization on micro-end milling of titanium alloy using nanofluid minimum quantity lubrication with chilly gas. *The International Journal of Advanced Manufacturing Technology*, 91(5), pp 2741–2749. Available at Doi: https://doi.org/10.1007/s00170-016-9965-6

Kumar, A., Ghosh, S. & Aravindan, S. 2017. Grinding performance improvement of silicon nitride ceramics by utilizing nanofluids. *Ceramics International*, 43(16), pp 13411–13421. Available at Doi: https://doi.org/10.1016/j.ceramint.2017.07.044

Lee, P.-H., Nam, T. S., Li, C. & Lee, S. W. Environmentally-friendly nano-fluid minimum quantity lubrication (MQL) meso-scale grinding process using nano-diamond particles. *2010 International Conference on Manufacturing Automation*, 2010. IEEE, 44–49. Available at Doi: https://doi.org/10.1109/ICMA.2010.27

Li, B., Li, C., Zhang, Y., Wang, Y., Yang, M., Jia, D., Zhang, N. & Wu, Q. 2017. Effect of the physical properties of different vegetable oil-based nanofluids on MQLC grinding temperature of Ni-based alloy. *The International Journal of Advanced Manufacturing Technology*, 89(9–12), pp 3459–3474. Available at Doi: https://doi.org/10.1007/s00170-016-9324-7

Li, K.-M. & Chou, S.-Y. 2010. Experimental evaluation of minimum quantity lubrication in near micro-milling. *Journal of Materials Processing Technology*, 210(15), pp 2163–2170. Available at Doi: https://doi.org/10.1016/j.jmatprotec.2010.07.031

Liao, Y. & Lin, H. 2007. Mechanism of minimum quantity lubrication in high-speed milling of hardened steel. *International Journal of Machine Tools and Manufacture*, 47(11), pp 1660–1666. Available at Doi: https://doi.org/10.1016/j.ijmachtools.2007.01.007

Minh, D.T., The, L.T. and Bao, N.T., 2017. Performance of Al2O3 nanofluids in minimum quantity lubrication in hard milling of 60Si2Mn steel using cemented carbide tools. *Advances in Mechanical Engineering*, 9(7), p.1687814017710618. Available at Doi: https://doi.org/10.1177%2F1687814017710618

Mosleh, M., Shirvani, K. A., Smith, S. T., Belk, J. H. & Lipczynski, G. 2019. A study of minimum quantity lubrication (MQL) by nanofluids in orbital drilling and tribological testing. *Journal Of Manufacturing And Materials Processing*, 3(1), pp 5. Available at Doi: https://doi.org/10.3390/jmmp3010005

Muhammad, J., Khan, A. M., Hussien, H., Gong, L., Mozammel, M., Gupta, M. K. & He, N. 2019. Effects of hybrid Al 2 O 3-CNT nanofluids and cryogenic cooling on machining of Ti–6Al–4V. *The International Journal of Advanced Manufacturing Technology*, 102(9–12), pp 3895–3909. Available at Doi: https://doi.org/10.1007/s00170-019-03485-9

Nam, J. S., Lee, P.-H. & Lee, S. W. 2011. Experimental characterization of microdrilling process using nanofluid minimum quantity lubrication. *International Journal of Machine Tools and Manufacture*, 51(7–8), pp 649–652. Available at Doi: https://doi.org/10.1016/j.ijmachtools.2011.04.005

Osman, K. A., Yılmaz, V., Ünver, H. Ö., Şeker, U. & Kılıç, S. E. 2020. Slot milling of titanium alloy with hexagonal boron nitride and minimum quantity lubrication and multi-objective process optimization for energy efficiency. *Journal of Cleaner Production*, 258, p. 120739. Available at Doi: https://doi.org/10.1016/j.jclepro.2020.120739

Padmini, R., Krishna, P. V. & Rao, G. K. M. 2016. Effectiveness of vegetable oil based nanofluids as potential cutting fluids in turning AISI 1040 steel. *Tribology International*, 94, pp 490–501. Available at Doi: https://doi.org/10.1016/j.triboint.2015.10.006

Pal, A., Chatha, S. S. & Sidhu, H. S. 2020a. Experimental investigation on the performance of MQL drilling of AISI 321 stainless steel using nano-graphene enhanced vegetable-oil-based cutting fluid. *Tribology International*, 151, p 106508. Available at Doi: https://doi.org/10.1016/j.triboint.2020.106508

Pal, A., Chatha, S. S. & Sidhu, H. S. 2021. Tribological characteristics and drilling performance of nano-MoS 2-enhanced vegetable oil-based cutting fluid using eco-friendly MQL technique in drilling of AISI 321 stainless steel. *Journal of the Brazilian Society of Mechanical Sciences and Engineering*, 43(4), pp 1–20. Available at Doi: https://doi.org/10.1007/s40430-021-02899-5

Pal, A., Chatha, S. S. & Singh, K. 2020b. Performance evaluation of minimum quantity lubrication technique in grinding of AISI 202 stainless steel using nano-MoS 2 with vegetable-based cutting fluid. *The International Journal of Advanced Manufacturing Technology*, 110(1), pp 125–137. Available at Doi: https://doi.org/10.1007/s00170-020-05840-7

Prabhu, K. N., Fernades, P. & Kumar, G. 2009. Effect of substrate surface roughness on wetting behaviour of vegetable oils. *Materials & Design*, 30(2), pp 297–305. Available at Doi: https://doi.org/10.1016/j.matdes.2008.04.067

Qingan, Y., Li, C., Zhang, Y., Yang, M., Dongzhou, J., Yali, H., Li, R. & Dong, L. 2018. Spectral analysis and power spectral density evaluation in Al 2 O 3 nanofluid minimum quantity lubrication milling of 45 steel. *The International Journal of Advanced Manufacturing Technology*, 97(1–4), pp 129–145. Available at Doi: https://doi.org/10.1007/s00170-018-1942-9

Rahim, E., Ibrahim, M., Rahim, A., Aziz, S. & Mohid, Z. 2015. Experimental investigation of minimum quantity lubrication (MQL) as a sustainable cooling technique. *Procedia CIRP*, 26, pp 351–354. Available at Doi: https://doi.org/10.1016/j.procir.2014.07.029

Sadeghi, M., Hadad, M., Tawakoli, T., Vesali, A. & Emami, M. 2010. An investigation on surface grinding of AISI 4140 hardened steel using minimum quantity lubrication-MQL technique. *International Journal of Material Forming*, 3(4), pp 241–251. Available at Doi: https://doi.org/10.1007/s12289-009-0678-3

Said, Z., Gupta, M., Hegab, H., Arora, N., Khan, A. M., Jamil, M. & Bellos, E. 2019. A comprehensive review on minimum quantity lubrication (MQL) in machining processes using nano-cutting fluids. *The International Journal of Advanced Manufacturing Technology*, 105(5), pp 2057–2086. Available at Doi: https://doi.org/10.1007/s00170-019-04382-x

Sen, B., Gupta, M. K., Mia, M., Mandal, U. K. & Mondal, S. P. 2020. Wear behaviour of TiAlN coated solid carbide end-mill under alumina enriched minimum quantity palm oil-based lubricating condition. *Tribology International*, 148, p 106310. Available at Doi: https://doi.org/10.1016/j.triboint.2020.106310

Setti, D., Sinha, M. K., Ghosh, S. & Rao, P. V. 2015. Performance evaluation of Ti–6Al–4V grinding using chip formation and coefficient of friction under the influence of nanofluids. *International Journal of Machine Tools and Manufacture*, 88, pp 237–248. Available at Doi: https://doi.org/10.1016/j.ijmachtools.2014.10.005

Shabgard, M., Seyedzavvar, M. & Mohammadpourfard, M. 2017. Experimental investigation into lubrication properties and mechanism of vegetable-based CuO nanofluid in MQL grinding. *The International Journal of Advanced Manufacturing Technology*, 92(9), pp 3807–3823. Available at Doi: https://doi.org/10.1007/s00170-017-0319-9

Shafi, W. K. & Charoo, M. 2018. Nanolubrication systems: an overview. *Materials Today: Proceedings*, 5(9), pp 20621–20630. Available at Doi: https://doi.org/10.1016/j.matpr.2018.06.443

Shafi, W. K. & Charoo, M. 2021a. An experimental study on the effect of concentration of green nanoadditives on the tribological properties of the biolubricants. *Proceedings of the Institution of Mechanical Engineers, Part C: Journal of Mechanical Engineering Science*, 09544062211042046. Available at Doi: https://doi.org/10.1177%2F09544062211042046

Shafi, W. K. & Charoo, M. 2021b. An overall review on the tribological, thermal and rheological properties of nanolubricants. *Tribology - Materials, Surfaces and & Interfaces*, 15(1), pp 20–54. Available at Doi: https://doi.org/10.1080/17515831.2020.1785233

Shafi, W. K., Charoo, M. & Hanief, M. 2021. Effect of fatty acid composition on the lubricating properties of bio-based green lubricants. *Tribology and Sustainability*. CRC Press. Available at ISBN: 9781003092162

Shafi, W. K., Raina, A. & Haq, M. I. U. 2018. Tribological performance of avocado oil containing copper nanoparticles in mixed and boundary lubrication regime. *Industrial Lubrication and Tribology* Available at Doi: https://doi.org/10.1108/ILT-06-2017-0166

Singh, H., Sharma, V. S. & Dogra, M. 2020a. Exploration of graphene assisted vegetables oil based minimum quantity lubrication for surface grinding of TI-6AL-4V-ELI. *Tribology International*, 144, p 106113. Available at Doi: https://doi.org/10.1016/j.triboint.2019.106113

Singh, P., Dureja, J., Singh, H. & Bhatti, M. 2019a. Nanofluid-based minimum quantity lubrication (MQL) face milling of inconel 625. *International Journal of Automotive and Mechanical Engineering*, 16(3), pp 6874–6888. Available at Doi: https://doi.org/10.15282/ijame.16.3.2019.04.0516

Singh, R., Dureja, J. & Dogra, M. 2019b. Performance evaluation of textured carbide tools under environment-friendly minimum quantity lubrication turning strategies. *Journal of the Brazilian Society of Mechanical Sciences and Engineering*, 41(2), p 87. Available at Doi: https://doi.org/10.1007/s40430-019-1586-1

Singh, R., Dureja, J., Dogra, M., Gupta, M. K., Mia, M. & Song, Q. 2020b. Wear behavior of textured tools under graphene-assisted minimum quantity lubrication system in machining Ti-6Al-4V alloy. *Tribology International*, 145, p 106183. Available at Doi: https://doi.org/10.1016/j.triboint.2020.106183

Singh, R., Dureja, J. S., Dogra, M., Gupta, M. K. & Mia, M. 2019c. Influence of graphene-enriched nanofluids and textured tool on machining behavior of Ti-6Al-4V alloy. *The International Journal of Advanced Manufacturing Technology*, 105(1), pp 1685–1697. Available at Doi: https://doi.org/10.1007/s00170-019-04377-8

Singh, T., Dureja, J. S., Dogra, M. & Bhatti, M. S. 2018. Environment friendly machining of Inconel 625 under nano-fluid minimum quantity lubrication (NMQL). *International Journal of Precision Engineering and Manufacturing*, 19(11), pp 1689–1697. Available at Doi: https://doi.org/10.1007/s12541-018-0196-7

Sreejith, P. & Ngoi, B. 2000. Dry machining: machining of the future. *Journal of Materials Processing Technology*, 101(1–3), pp 287–291. Available at Doi: https://doi.org/10.1016/S0924-0136(00)00445-3

Su, Y., Gong, L., Li, B., Liu, Z. & Chen, D. 2016. Performance evaluation of nanofluid MQL with vegetable-based oil and ester oil as base fluids in turning. *The International Journal of Advanced Manufacturing Technology*, 83(9–12), pp 2083–2089. Available at Doi: https://doi.org/10.1007/s00170-015-7730-x

Sugiantoro, B. 2018. Performance evaluation of using water and bio oil-based nanocutting fluids under minimum quantity lubrication with compressed cold air during milling operations of steel. *IOP Conference Series: Materials Science and Engineering*, p 012098. Available at Doi: https://doi.org/10.1088/1757-899X/403/1/012098

Talon, A. G., Lopes, J. C., Sato, B. K., Tavares, A. B., Ribeiro, F. S. F., Genovez, M. C., Pinto, T. A. D., de Mello, H. J., Aguiar, P. R. & Bianchi, E. C. 2020. Grinding performance of hardened steel: a study about the application of different cutting fluids with corrosion inhibitor. *The International Journal of Advanced Manufacturing Technology*, 108(9), pp 2741–2754. Available at Doi: https://doi.org/10.1007/s00170-020-05598-y

Tawakoli, T., Hadad, M., Sadeghi, M., Daneshi, A., Stöckert, S. & Rasifard, A. 2009. An experimental investigation of the effects of workpiece and grinding parameters on minimum quantity lubrication—MQL grinding. *International Journal of Machine Tools and Manufacture*, 49(12–13), pp 924–932. Available at Doi: https://doi.org/10.1016/j.ijmachtools.2009.06.015

Uysal, A., Demiren, F. & Altan, E. 2015. Applying minimum quantity lubrication (MQL) method on milling of martensitic stainless steel by using nano MoS2 reinforced vegetable cutting fluid. *Procedia-Social and Behavioral Sciences*, 195, pp 2742–2747. Available at Doi: https://doi.org/10.1016/j.sbspro.2015.06.384

Virdi, R. L., Chatha, S. S. & Singh, H. 2021. Performance evaluation of nanofluid-based minimum quantity lubrication grinding of Ni-Cr alloy under the influence of CuO nanoparticles. *Advances in Manufacturing*, 9(4), pp 580–591. Available at Doi: https://doi.org/10.1007/s40436-021-00362-1

Wang, Y., Li, C., Zhang, Y., Li, B., Yang, M., Zhang, X., Guo, S. & Liu, G. 2016. Experimental evaluation of the lubrication properties of the wheel/workpiece interface in MQL grinding with different nanofluids. *Tribology International*, 99, pp 198–210. Available at Doi: https://doi.org/10.1016/j.triboint.2016.03.023

Yin, Q., Li, C., Dong, L., Bai, X., Zhang, Y., Yang, M., Jia, D., Hou, Y., Liu, Y. & Li, R. 2018. Effects of the physicochemical properties of different nanoparticles on lubrication performance and experimental evaluation in the NMQL milling of Ti–6Al–4V. *The International Journal of Advanced Manufacturing Technology*, 99(9), pp 3091–3109. Available at Doi: https://doi.org/10.1007/s00170-018-2611-8

Yuan, S., Hou, X., Wang, L. & Chen, B. 2018. Experimental investigation on the compatibility of nanoparticles with vegetable oils for nanofluid minimum quantity lubrication machining. *Tribology Letters*, 66(3), pp 1–10. Available at Doi: https://doi.org/10.1007/s11249-018-1059-1

Zhang, D., Li, C., Jia, D., Zhang, Y. & Zhang, X. 2015a. Specific grinding energy and surface roughness of nanoparticle jet minimum quantity lubrication in grinding. *Chinese Journal of Aeronautics*, 28(2), pp 570–581. Available at Doi: https://doi.org/10.1016/j.cja.2014.12.035

Zhang, X., Li, C., Jia, D., Gao, T., Zhang, Y., Yang, M., Li, R., Han, Z. & Ji, H. 2019. Spraying parameter optimization and microtopography evaluation in nanofluid minimum quantity lubrication grinding. *International Journal of Advanced Manufacturing Technology*, 103 (5–8), pp 2523–2539. Available at Doi: https://doi.org/10.1007/s00170-019-03642-0

Zhang, Y., Li, C., Jia, D., Li, B., Wang, Y., Yang, M., Hou, Y. & Zhang, X. 2016a. Experimental study on the effect of nanoparticle concentration on the lubricating property of nanofluids for MQL grinding of Ni-based alloy. *Journal of Materials Processing Technology*, 232, pp 100–115. Available at Doi: https://doi.org/10.1016/j.jmatprotec.2016.01.031

Zhang, Y., Li, C., Jia, D., Zhang, D. & Zhang, X. 2015b. Experimental evaluation of MoS2 nanoparticles in jet MQL grinding with different types of vegetable oil as base oil. *Journal of Cleaner Production*, 87, pp 930–940. Available at Doi: https://doi.org/10.1016/j.jclepro.2014.10.027

Zhang, Y., Li, C., Yang, M., Jia, D., Wang, Y., Li, B., Hou, Y., Zhang, N. & Wu, Q. 2016b. Experimental evaluation of cooling performance by friction coefficient and specific friction energy in nanofluid minimum quantity lubrication grinding with different types of vegetable oil. *Journal of Cleaner Production*, 139, pp 685–705. Available at Doi: https://doi.org/10.1016/j.jclepro.2016.08.073

Chapter 6

Tribological performance and microstructural analysis of Al-Si/coconut shell ash metal matrix composites under varied reinforcement content and lubrication conditions

Sivasankara Raju Rallabandi, Gunji Srinivasa Rao, Chennupati Lakshmi Srinivas, and Durga Prasad Kelli

6.1 INTRODUCTION: BACKGROUND AND DRIVING FORCES

Composite materials are multiphase materials that exhibit both constituent phases' features on a significant scale, resulting in a finer integration of qualities. This law of cooperative endeavor states that intentional mixing of two or more dissimilar materials results in further property combinations. "A multiphase material presented in which the component phases must be disparate and distributed by a discrete interface" is the conventional description of the composite. Metal matrix composites (MMCs) have good performance in a variety of areas, including wear resistance, corrosive conditions, growth, stiffness, and hot pressure tolerance. Because of their improved features, MMCs are being evaluated for many different applications. Aluminum-based metal matrix composites are progressively emerging as viable substitute materials for connecting rods, cylinder blocks, gears, shafts, and other components because of all these advantages (Thimothya et al. 2019; Vuddagiri et al. 2021).

Aluminum and aluminum alloys are lightweight non-ferrous metals that are renowned for their exceptional ductility, strength, and resistance to attrition. When mixed with the parts of titanium/iron alloys, aluminum is a great conductor of both high energy and high temperatures. Aluminum compounds become excessively active because of titanium's hardness and aluminum's low weight (Vuddagiri et al. 2022). Titanium is just as hard as aluminum. Aluminum alloys are quite robust and stiff, which enhances certain modules and decreases stickiness. However, it is important to replace current materials (i.e., cast iron and steel) with aluminum metal matrix composites reinforced with ceramic particles in order to attain superior specific strength for future generation automotives and other industrial components. For tribological applications, cast aluminum metal matrix particle composites offer an inexpensive substitute for traditional materials (Rao et al. 2021).

DOI: 10.1201/9781003717362-6

The protective film is that lubricants provide generally results in more favorable wear characteristics under lubricated conditions as opposed to arid conditions. The particular application, the surrounding conditions, and the desired level of wear resistance all influence the decision between lubricated and dry conditions. Tribological studies are frequently carried out by researchers and engineers to identify the best lubricating techniques and materials to maximize the longevity and performance of mechanical components (Chen, Bosse, and Deters 2009).

Dry conditions often result in increased friction and wear in the absence of lubrication. The lack of a lubricant film increases the probability of metal-to-metal contact, which subsequently results in heightened levels of wear and friction. A significant feature of tribological investigations is the wear behavior in lubricated environments and its comparison with dry ones. A protective barrier is established by lubricants, including oils and greases, between rotating and sliding surfaces. This lessens wear and friction by reducing direct metal-to-metal contact. Certain lubricants are better suited for high-load, low-speed situations, while others are made for extremely high pressure. Lubricants aid in the dissipation of heat produced while rolling or sliding, avoiding localized overheating, which may cause failure, wear, and degradation of the material.

According to reports, most lubricants are petroleum-based and end up in the environment, where they contribute to an increase in toxicity. Vegetable oil-based lubricants are increasingly being used as an alternative to petroleum-based lubricants in order to address the environmental problem. Natural oil has a structure called triacylglycerol, which is crucial for lubrication because it has long chains of polar fatty acids that stick to metal surfaces (Biswas et al. 2022).

Numerous studies have focused on different vegetable oils, such as rapeseed, canola, soy, sunflower, and rice bran oil, which are used to make biodiesel-based lubricants. Some researchers have determined that while methyl esters formed from fatty acids are beneficial in reducing friction, individual fatty acid esters are not as successful in their research on biodiesel-based lubricants and additives used on it (Fox, Tyrer, and Stachowiak 2004). Research is ongoing on bio-esters and their effects on engines as lubricants and diesel, but in order to develop the possibility of real-time applications for both the materials and the bio-lubricants, it is also necessary to gain a clear understanding of how these compounds behave when they come into contact with newly developed materials.

Water is a problematic lubricant to utilize in many practical applications. One synthetic petroleum-based lubricant available commercially, one lubricant free of chlorofluorocarbon derived from petroleum oil blended with natural oil, and one bio-lubricant derived from vegetable oil are utilized in this study to determine the tribological parameters and contact surface behavior of stir-cast Al-TiB2 composites under lubricated conditions (Prasad and Asthana 2004). A lot of work has been done to characterize bio-lubricants, but not much has been done to look at how different bio-diesel-contaminated bio-lubricants affect the touch surface. Apart from the two lubricants that are commercially available, methyl ester derived from rice bran oil has been selected for development and application as a bio-lubricant in this study. This is because rice bran oil has lower wear resistance than other vegetable oils due to the presence of natural antioxidants such as tocopherols and gamma oryzanol. Rice bran oil stays stable at high temperatures up to 395 °C without losing any of its mass, and methyl esters are better at reducing friction (Sinha, Agarwal, and Garg 2008).

Chen et al. (2009) examine the wear and friction properties of several base oils and additives in lubricated steel-on-aluminum and aluminum-on-aluminum contacts. They discovered that polar base fluids, including monoester, polyolester, and polyalkylene glycol, showed inferior tribological properties compared to hydrocarbon-based fluids like mineral oil (MO) or polyalphaolefin (PAO). In addition, when applied to steel-on-aluminum contacts, a lubricant containing MO and zinc dialkyldithiophosphate (ZDDP) outperformed MO alone or MO combined with a phosphorus/sulfur additive in terms of reduced friction and improved wear resistance.

In this study, three different lubricants are used to conduct pin-on-disc tribological testing on stir-cast Al-CSA composites at varying speeds. In this study, SAE20W-40 and rice bran oil are used as lubricants to determine the reason the tribological performance of Al-Si/coconut shell ash metal matrix composites changes when using different lubricants. The worn surfaces in lubricated contact are characterized using scanning electron microscopy (SEM) pictures collected after tribological tests, and the tribological features of Al-CSA during lubricated sliding are compared to those observed during dry contact.

6.2 EXPERIMENTAL PROCEDURE

6.2.1 Al-Si as material and fabrications technique

Al-Si alloys differ from phase diagrams in that silicon has no solid solubility for aluminum. Because there is not a real-world issue that needs to be fixed, there is not a beta procedure. Al-Si alloys have excellent castability, strong corrosion resistance, and a very low density. The inability of alloys to undergo heat treatment is their main disadvantage. In order to improve the alloy's engineering qualities, magnesium is added to it. The stir casting method involves mixing ceramic particles into a liquid, melting it, and then solidifying it. A key component of this strategy is excellent wettability between the melt and the reinforcement. The figure illustrates the most straightforward, efficient, and user-friendly method, which is the stir-casting or vortex technique. The vortex system involved introducing pre-treated ceramic particles into a molten alloy via a chute, with a revolving impeller generating a vortex. Up to 30% CSA particles should be added to the melt. The hot slurry contains ceramic particles and a matrix that are transferred to a heated mold to take on the appropriate shape and allow the slurry to solidify. There have been reports of a thorough explanation elsewhere (Table 6.1).

6.2.2 Lubricants used

In this study, two distinct liquid lubricants are employed. SAE20W-40, an industrial lubricant made from petroleum, can be bought from Indian Oil Corporation Limited. This commercial lubricant has a well-balanced blend of additives and is based on highly refined petroleum stocks. Researchers have utilized lubricants of a similar nature in the past in order to learn more about the boundary lubricated tribology of an aluminum–silicon alloy sliding against steel (Das and Biswas 2004).

One bio-lubricant, called rice bran oil-based TMP ester, is made in a laboratory following the normal steps. The unrefined rice bran oil is procured from the many industrial establishments located in Srikakulam. After heating the crude oil to 110 °C, it is combined with a methanolic solution that has also been made by correctly combining KOH and methanol in a predetermined molar ratio. A motorized mixer is used to keep the processed crude oil and methanolic process moving while they are mixed. The combination forms two distinct layers after a day at 70 °C, with ester in the upper layer. We refer to this procedure as trans-esterification. For this investigation, the upper ester layer is extracted from the mixture, collected, and utilized as a bio-lubricant (Table 6.2).

Table 6.1 Composition of Al-Si

Fe	*Si*	*Cu*	*Zn*	*Mn*	*Mg*	*Bal.*
0–0.75%	0–0.85%	0.05–0.2%	0–0.1%	0–0.05%	0–0.15%	Al

Table 6.2 Properties of SAE20W-40 and rice bran oil

Lubricant	*Viscosity*	*Specific gravity*	*Flash point*	*Reference*
SAE20W-40	13.5–15.5cSt at 100 °C (ASTM D445)	0.8–0.9 g/ml	132	Das and Biswas (2004)
Rice bran oil-based TMP ester	4–5.5cSt at 40 °C	0.877 at 30 °C	183 °C	Poria, Sutradhar, and Sahoo (2016)

6.2.3 Determination techniques of mechanical properties

The cast composite's hardness was measured in accordance with ASTM-384 using a Vickers hardness tester with a 100-g (0.98N) applied stress. Each specimen had five indentations counted and averaged. A Hounsfield tensometer (ETM-ER3/772/12) reading at a strain rate is 5 × 10^4 s^{-1} revealed tensile characteristics. Under dry and soaking wet conditions, the sliding wear resistance of cast composites is determined using a DUCOM pin-on-disc tester (Model: TR-201 LE-PHM 400). Figure 6.1 shows a schematic design of the pin-on-disc setup. The test pieces, which are 8 mm in diameter and 27 mm long, were made according to the ASTM G99-95 guidelines.

Wear is when a piece of material loses its shape over a moving distance. A linear variable differential transformer (LVDT) is used to record the amount of material sliding in µm (Raju, Srinivasa Rao, and Samantra 2019). A measurement of the sliding can be made with ±2 mm precision with a least count of 0.1 µm. Volume loss has been produced from the weight decrease. The sliding rate can be determined by calculating the volume loss to sliding distance ratio (Rao and Das 2011) or by calculating the amount of material lost per unit time of abrasion. The wear rate is expressed in mm^3/m (Radhika 2017; Raju and Venkata Siva 2021). Frictional forces per unit normal load have also been used to compute the coefficient of friction (COF). To improve

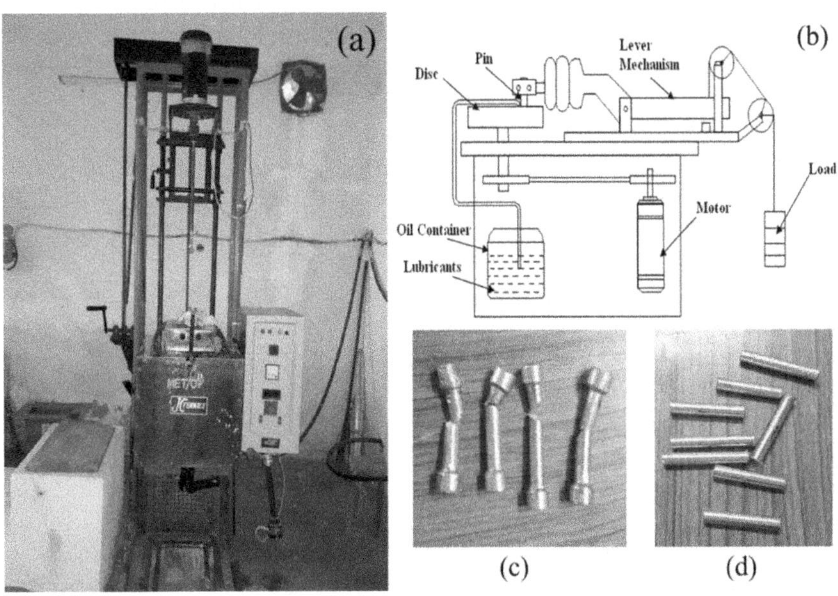

Figure 6.1 (a) Stir casting set-up (b) pin-on-disc set-up (c) tensile samples and (d) cylindrical wear samples.

the outcome, the COF mean has been taken into account. Furthermore, the wear tracks of the materials are examined using scanning electron microscopy (SEM).

6.3 RESULTS AND DISCUSSIONS

6.3.1 Properties of CSA composites

As shown in Figure 6.2, the density of CSA composites declined by 1.85, 3.69, and 5.90% when CSA content was 5%, 10%, and 15%, respectively, when compared to Al-Si alloy. It was found that the hardness of Al-CSA composites went up as the CSA weight went up by up to 15%. Composites with 5, 10, and 15% CSA are more durable than their Al-Si alloy counterparts by 6.71%, 36.69%, and 42.87%, respectively. The presence of alkali earth metals like CaO and MgO in CSA contributed to the improved hardness of the composite, as did the uniform distribution of CSA in the material. As seen in Figure 6.2, the tensile strength rose as the weight percentage of CSA particles increased by up to 15%. In addition, elongation decreases as CSA particle addition increases. Even though the elongation decreases as the amount of CSA particles added increases, the tensile strength improves as the volume percentage of CSA particles increases. The addition of hard ceramic-reinforced particles has resulted in an increase in the strength of composite materials. The mechanisms of grain size are strengthened by the contacts between the matrix and reinforcement. Particle movement has been crucial, and reinforced CSA particles improve load-bearing capacity and limit matrix dislocation movement to minimize interspacing (Alpas and Zhang 1994; Rallabandi et al. 2019).

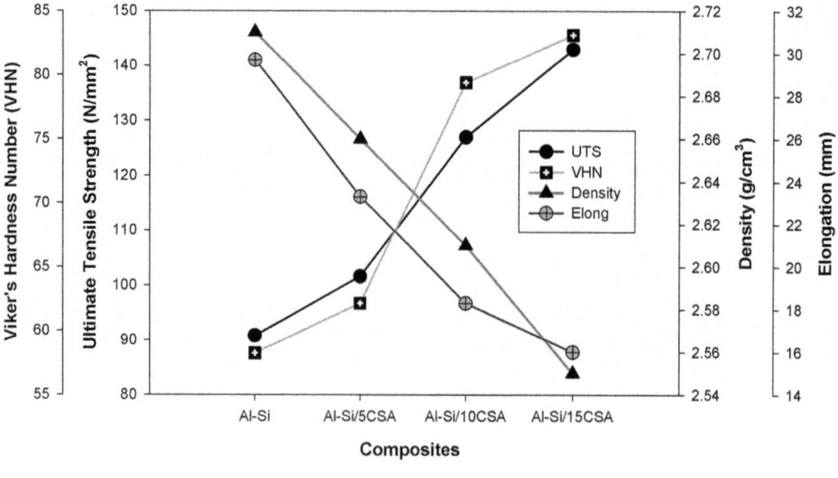

Figure 6.2 Physical and mechanical properties of CSA composites.

6.3.2 Wear behavior in lubricated conditions compared with dry conditions

As shown in Figure 6.3, the wear test results of Al-CSA composites with different lubricants were done with a 30 N load on the pin and a 3000 m moving distance. The wear results for three distinct sliding velocities and weight percentages of Al-CSA composites, including base alloy with two lubricants, such as SAE 40W-20 and rice bran oil-based TMP ester, are shown in the figure. Regardless of the kind of lubricant, wear increases with increasing sliding velocity for all CSA % composites, and this pattern is consistent with Archard's wear rule (Archard 1953). In composites, wear reduces as the weight percentage of CSA rises. This is because the composite's increased CSA content causes an increase in hardness. Because the Al-Si matrix has a low stacking defect energy, work hardening occurs more quickly. Because composites are work-hardened, there is an initial period of difficulty and energy-consuming moments between tribo surfaces (i.e., pin and disc), which increases frictional force and plastic deformation. In terms of wear reduction, rice bran oil performs about identically to SAE at a certain 30 N load and speed combination. The difference in wear between the base alloy and the 5% CSA composite is approximately 13% and 6%, respectively, for the same load, or 30 N. Similar patterns can be seen in Figure 6.3, which shows the wear results for composites with varying weight percentages and loads using commercial lubricant SAE20W-40 and rice bran oil. When it comes to wear resistance, commercial SAE20W-40 is superior, with rice bran oil coming in second. All of the sliding velocity instances in Figure 6.3 follow this trend.

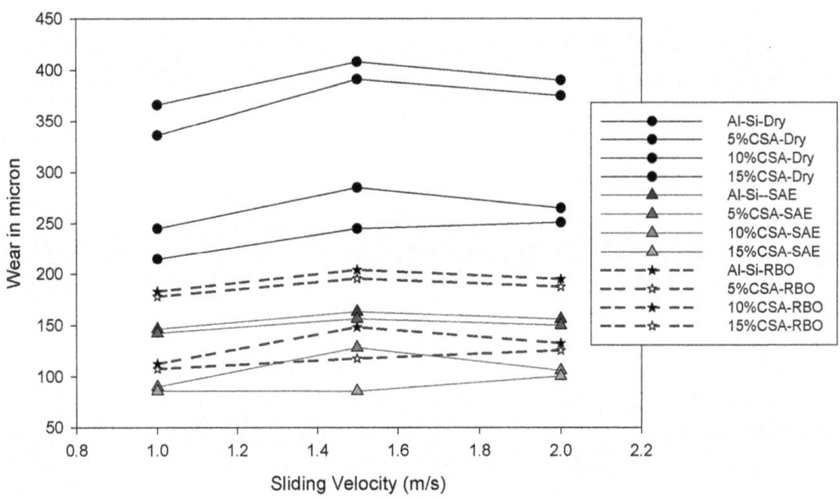

Figure 6.3 Wear among weight fraction of CSA for varying speeds at a constant load of 30 N, sliding distance of 3000 m for SAE20W-40 and Rice bran oil.

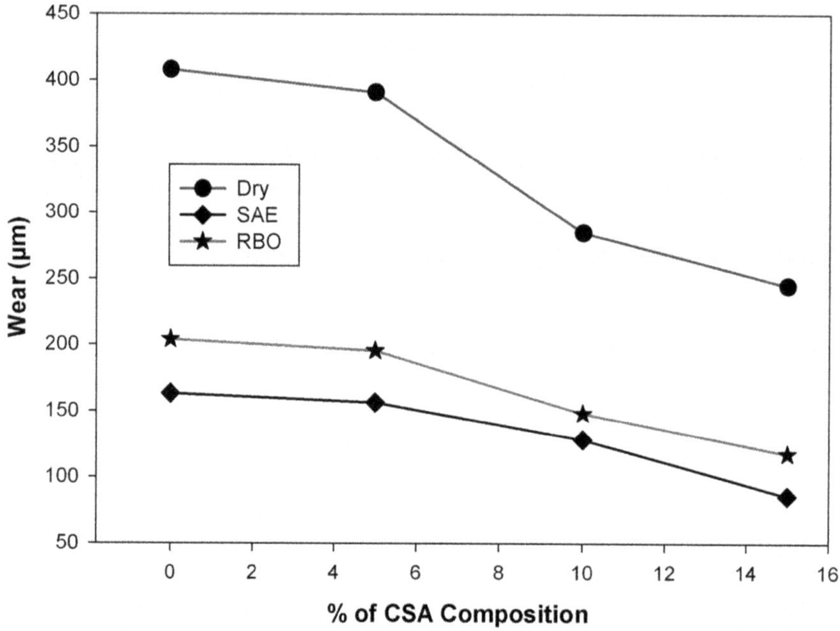

Figure 6.4 Wear among weight percentages of CSA under dry and lubricated conditions.

Figure 6.4 illustrates the effects of two commercial lubricants, that is, one containing bio-additives and the other containing petroleum base and dry sliding on the wear of various weight fractions of CSA. As the weight percent of CSA increases, wear diminishes, and the effects of the two lubricants are nearly identical in nature, in accordance with dry sliding trends, under all conditions of 30 N load, 3000 m sliding distance, and 1.5 m/s sliding velocity. Industrial SAE20W-40 is the best wear-resistant lubricant for the composite steel contact interface, as shown in Figure 6.4.

6.3.3 Friction behavior in lubricated conditions compared with dry conditions

Figure 6.5 illustrates the impact of motion on the tribological response of Al-CSA composites, specifically the coefficient of friction, in both dry and lubricated conditions. The graph in concern depicts a comparison between the coefficient of friction and the weight percent of CSA for RBO-based TMP ester SAE20W-40. It has been observed that the coefficient of friction reduces with an increase in the weight percent of the CSA as well as an increase in the sliding velocity under all situations, including dry and lubricated. The base matrix exhibits a reduced coefficient of friction under arid conditions owing to its soft composition. The reduced coefficient of friction can be attributed to the reinforced particles that create a mechanically

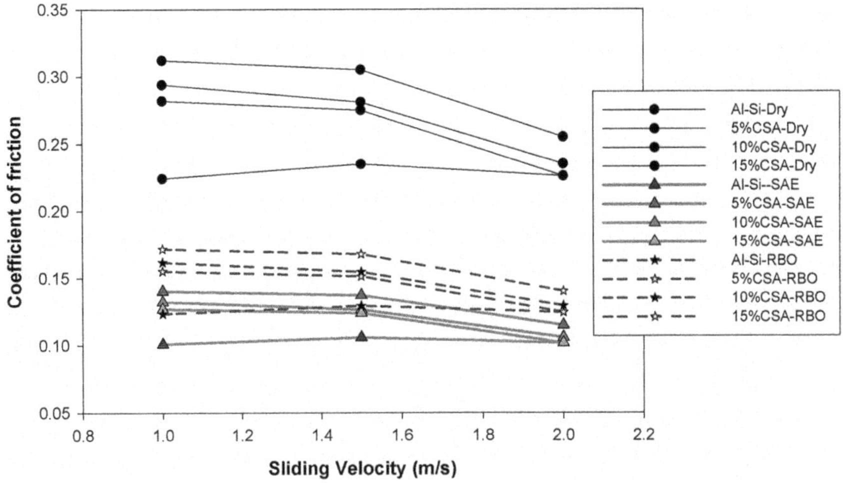

Figure 6.5 Coefficient of friction among CSA composites for varying velocity in comparison of dry, rice bran oil and SAE20W-40.

mixed layer at the contact surface, providing protection as the sliding speed increases. As a result, there is less wear and friction. Wear and the coefficient of friction are the two tribological performance metrics that are negatively correlated with the volume of CSA (Siva Sankara Raju et al. 2017). For RBO TMP ester lubricant, friction coefficient values are in the range between 0.121 and 0.171 for a 30 N load. For base alloy and 5%CSA, the change in friction values for the same lubricant with sliding velocities from 1 m/s to 2 m/s is approximately 17% and 20%, respectively. In RBO lubricant, the friction values for 5% CSA fall between 0.140 and 0.172. Likewise, the SAE20W-40 lubricant falls within the range of 0.114–0.140. It can be observed that friction values for dry lubricant fluctuate more with sliding velocity than for the other two lubricants, and for commercial lubricant, friction values more closely fall within the same range. Friction values are lower for RBO lubricant at the same composite weight percentage and velocity, and they are significantly lower at higher velocities than at higher weight percentages. The best lubricant for the frictional behavior of Al-CSA composites seems to be RBO. From an application standpoint, the lowest value of friction found under these parametric conditions is 0.101, and this value remains constant when RBO lubricant is present.

In lubricated situations, frictional force occurs owing to shear force generation inside the lubrication layer, and this shear force depends on lubricant properties. Consequently, lubricants possessing a closed structure will exhibit comparable frictional characteristics (Sinha, Agarwal, and Garg 2008). This results in nearly identical friction values for Al-CSA composites when two commercial lubricants, RBO and SAE, are used; yet, these provide distinct wear values, as seen in Figure 6.5. This kind of action is seen in lubricants that are not absorbed by the surface; on the other hand, lubricants that are absorbed by the metallic surface generate a metallic soap that prevents asperities from approaching and lowers frictional values. The capacity to produce metallic soap and reduce friction is independent of contact loads. Within specific parametric values, the action of metallic soap on the surface exceeds the produced shear force. Because methyl esters can be absorbed to create metallic soap, there are, hence, reduced frictional values when they are present on the contact surface (Kato 2000; Jeyaprakash and Yang 2020).

Figure 6.6 compares the friction in greased and dry conditions. It is evident that weight percentages of CSA have a considerable impact on friction in dry conditions, whereas in lubricated conditions, lines are nearly parallel to the axis. This indicates that the lubricant's impact on friction is significantly more than that of the CSA's weight percentage. With 10% CSA composites, friction values in a dry condition range from 0.235 to 0.294 with variable velocity; with SAE lubrication, these values range from 0.105 to 0.132. In actuality, lubricated conditions have twice as much friction as dry conditions. It is clear that lubricants play a critical role in reducing friction, a role that cannot be made up for by adding additional CSA particles to weight percentages as reinforcement. For RBO lubricant, the difference between the friction values in the lubricated and dry states is likewise greater. Higher velocities, however, could ruin this situation, and it will be challenging to accomplish in the presence of lubricants of the RBO ester type. The research on wear and friction in metal matrix composites has gained more

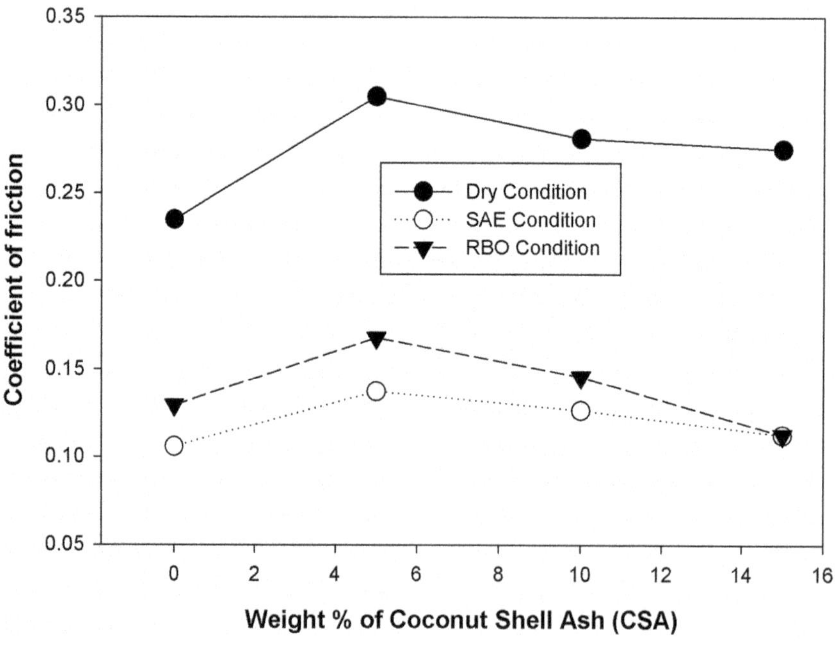

Figure 6.6 Comparison of coefficient friction among CSA composites in dry and lubricated conditions.

attention recently (Raju, Murali, and Patnaik 2020). All these investigations, however, only take into account dry contact scenarios. Thus, the current investigation will contribute to expanding the application of these composites under different lubricating circumstances.

6.4 SEM CHARACTERIZATION OF WORN-OUT SURFACES UNDER SAE LUBRICANT

In order to analyze the mechanisms of wear during lubricated contact sliding, Figure 6.7 displays SEM micrographs of worn surfaces in SAE20W-40 lubricated contacts operating at 30 N and 1.5 m/s. The appearance of longitudinal grooves, partial pits, micro-cutting, and micro plowing under dry conditions suggests that both adhesive and abrasion wear mechanisms are involved.

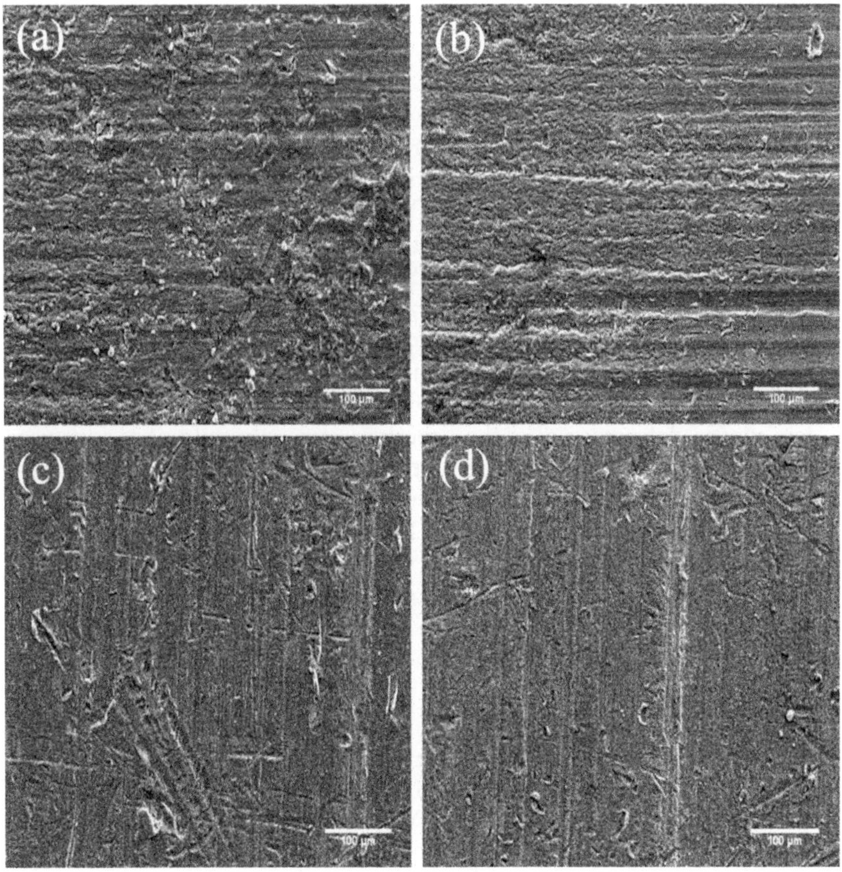

Figure 6.7 SEM images of worn surfaces of CSA composites (a) Al-Si alloy (b) 5%CSA (c) 10%CSA (d) 15%CSA.

However, in lubricated conditions, virtually few signs of plowing, cracking, or pitting are seen on the worn surfaces, with the exception of 10% and 15% CSA composites. Thus, the lubricants' corrosiveness has no effect. Asperity elastic deformation is thought to be a key mechanism for the composite surface under lubrication. The surface experiences less plowing as a result of wear brought on by elastic deformation. Higher CSA weight percentages cause the composite to become harder and therefore increase the likelihood of surface abrasion during sliding. This phenomenon is seen when the composite is dry, but in lubricated conditions, higher CSA weight percentages cause harder asperities at the mating surfaces, which reduces the likelihood of plastic deformation. Hard particles are formed and placed on the surface under dry conditions due to the plastic deformation of asperities; however, in lubricated conditions, plastic deformation is limited to 5%CSA composites, resulting in a somewhat rough surface. Smoother surfaces are a result of lubricated conditions since lubricants also aid in wiping out the tips that are acquired during sliding or rolling motion.

6.5 CONCLUSION

In this study, Al-CSA composites are produced using the stir casting method, and tribological experiments are done in a pin-on-disc triobotester with two different lubricants. With a rise in CSA weight fraction, the composite's properties, such as tensile strength, elasticity of modulus, and hardness, are increased. As the weight fraction of CSA increased, the density and elongation decreased. When asperities slide across the steel surface in lubricated contact, they predominantly flex elastically because of the larger weight percentage of 15%CSA composites. It results in wear trails with a comparatively smoother surface morphology, as confirmed by SEM images. Because there is less plastic deformation of the asperities and only mild abrasion wear, there is less plowing and cracking on the surface of the 5%CSA composite. Commercial lubricants with a petroleum base are the most wear-resistant, and they are followed in effectiveness by lubricants with bio-additives. Bio-based lubricants are the most effective when it comes to frictional properties because they can partially produce metallic soap through surface absorption.

REFERENCES

Alpas, A. T., and J. Zhang. 1994. "Effect of Microstructure (Particulate Size and Volume Fraction) and Counterface Material on the Sliding Wear Resistance of Particulate-Reinforced Aluminum Matrix Composites." *Metallurgical and Materials Transactions A* 25 (5): 969–83. https://doi.org/10.1007/BF02652272

Archard, J. F. 1953. "Contact and Rubbing of Flat Surfaces." *Journal of Applied Physics* 24 (8): 981–88. https://doi.org/10.1063/1.1721448

Biswas, Md Abu Sayeed, Md Mashfiqur Rahman, Javier A. Ortega, Laura Peña-Parás, Demófilo Maldonado-Cortés, José A. González, Ricardo Cantú, Adrián Campos, and Eugenio Flores. 2022. "Lubrication Performance of Sunflower Oil Reinforced with Halloysite Clay Nanotubes (HNT) as Lubricant Additives." *Lubricants* 10 (7): 139. https://doi.org/10.3390/lubricants10070139

Chen, C., H. Bosse, and L. Deters. 2009. "Effects of Various Base Oils and Additives on the Tribological Behaviour of Lubricated Aluminium- on-Aluminium and Steel-on-Aluminium Contacts." *Proceedings of the Institution of Mechanical Engineers, Part J: Journal of Engineering Tribology* 223 (3): 571–80. https://doi.org/10.1243/13506501JET511

Das, S., and S. Biswas. 2004. "Boundary Lubricated Tribology of an Aluminium-Silicon Alloy Sliding Against Steel." *Tribology Letters* 17: 623–628. https://doi.org/10.1023/B:TRIL.0000044512.48967.02

Fox, N. J., B. Tyrer, and G. W. Stachowiak. 2004. "Boundary Lubrication Performance of Free Fatty Acids in Sunflower Oil." *Tribology Letters* 16 (4): 275–81. https://doi.org/10.1023/B:TRIL.0000015203.08570.82

Jeyaprakash, Natarajan, and Che-Hua Yang. 2020. "Friction, Lubrication, and Wear." In *Tribology in Materials and Manufacturing*, edited by Amar Patnaik, Tej Singh, and Vikas Kukshal, 1st ed., 326–42. London, SW7 2QJ, UK: IntechOpen. https://doi.org/10.5772/INTECHOPEN.93796

Kato, Koji. 2000. "Wear in Relation to Friction—A Review." *Wear* 241: 151–57. https://doi.org/10.1016/s0043-1648(00)00382-3

Poria, S., G. Sutradhar, and P. Sahoo. 2016. "Wear and Friction Behavior of Stir Cast Al-TiB2 Metal Matrix Composites with Various Lubricants." *Tribology in Industry* 38 (4): 508–21.

Prasad, S. V., and R. Asthana. 2004. "Aluminum Metal – Matrix Composites for Automotive Applications: Tribological Considerations." *Tribology Letters* 17 (3): 445–53. https://doi.org/10.1023/B:TRIL.0000044492.91991.f3

Radhika, N. 2017. "Fabrication of LM25 / SiO 2 Metal Matrix Composite and Optimization of Wear Process Parameters Using Design of Experiment." *Tribology in Industry* 39 (1): 1–8. https://doi.org/10.24874/ti.2017.39.01.01

Raju, R. Siva Sankara, and B. Venkata Siva. 2021. "Fabrication and Tribological Studies of Al-CSA Composite Using RSM." *International Journal of Materials Engineering Innovation* 12 (2): 83–102. https://doi.org/10.1504/IJMATEI.2021.115596

Raju, Siva S., Gunji B. Murali, and Prabina K. Patnaik. 2020. "Ranking of Al-CSA Composite by MCDM Approach Using AHP–TOPSIS and MOORA Methods." *Journal of Reinforced Plastics and Composites* 39 (19–20): 721–32. https://doi.org/10.1177/0731684420924833

Rallabandi, Sivasankara Raju, Gunji Srinivasa Rao, R. Siva Sankara Raju, G. Srinivasa Rao, Sivasankara Raju Rallabandi, and Gunji Srinivasa Rao. 2019. "Assessment of Tribological Performance of Al-Coconut Shell Ash Particulate—MMCs Using Grey-Fuzzy Approach." *Journal of The Institution of Engineers (India): Series C* 100 (1): 13–22. https://doi.org/10.1007/s40032-017-0388-4

Rao, R. N., and S. Das. 2011. "Effect of SiC Content and Sliding Speed on the Wear Behaviour of Aluminium Matrix Composites." *Materials and Design* 32 (2): 1066–71. https://doi.org/10.1016/j.matdes.2010.06.047

Rao, C. J., R. Siva Sankara Raju, D. Sreeramulu, and Y. Sagar. 2021. "Fabrication and Evaluation of Optimized Parametric Condition During EDM Machining of Al-CSA Composite Using Taguchi Orthogonal Array." In *Advances in Design and Thermal Systems: Select Proceedings of ETDMMT 2020*, edited by Ganippa Lionel, R Karthikeyan, and V Muralidharan, 1st ed., 411–20. Springer Singapore. https://doi.org/10.1007/978-981-33-6428-8_33

Sinha, Shailendra, Avinash Kumar Agarwal, and Sanjeev Garg. 2008. "Biodiesel Development from Rice Bran Oil: Transesterification Process Optimization and Fuel Characterization." *Energy Conversion and Management* 49 (5): 1248–57. https://doi.org/10.1016/j.enconman.2007.08.010

Siva Sankara Raju, R., M. K. Panigrahi, R. I. Ganguly, and G. Srinivasa Rao. 2017. "Investigation of Tribological Behavior of a Novel Hybrid Composite Prepared with Al-Coconut Shell Ash Mixed with Graphite." *Metallurgical and Materials Transactions A* 48 (8): 3892–3903. https://doi.org/10.1007/s11661-017-4139-1

Siva Sankara Raju, R., G. Srinivasa Rao, and Chitrasen Samantra. 2019. "Wear Behavioral Assessment of Al-CSAp-MMCs Using Grey-Fuzzy Approach." *Measurement* 140 (July): 254–68. https://doi.org/10.1016/j.measurement.2019.04.004

Thimothya, P., S. Sankara Raju, C. Ratnam. 2019. "Development and Accretion of Tribological Performance on Al-CSA Composites Using Orthogonal Array." *Materials Today: Proceedings* 18 (7): 5332–39. https://doi.org/10.1016/j.matpr.2019.07.558

Vuddagiri, Hari Kiran, Srinivas Vadapalli, Jaikumar Sagari, and R. Sivasankara Raju. 2021. "Fabrication and Modelling of Tribological Performance of Al-Si/12Al2O3/2MoS2 Composite Using Taguchi Technique." *International Journal of Automotive and Mechanical Engineering* 18 (3): 8959–77. https://doi.org/10.15282/ijame.18.3.2021.09.0686

Vuddagiri, Hari Kiran, Sivasankara Raju Rallabandi, Srinivas Vadapalli, and Thimothy Pandi. 2022. "Assessment of Mechanical and Tribological Performance of Hybrid Al/MoS2/Al2O3 Composite by GFRA." *Metallurgical and Materials Engineering* 28 (1): 79–102. https://doi.org/10.30544/764

Chapter 7

Understanding oral tribology

Health concern for next generation

Roopsandeep Bammidi, Sreeramulu Dowluru, Kelli Durgaprasad, Hymavathi Madivada, Venkatesh Muddada, and Grandhi Prasanth

7.1 INTRODUCTION

Bio-tribology, which examines the connections between tribology and biology while taking into account biological materials and their use in medicine, is one of the application fields that has grown significantly in recent years. Through the introduction of various new materials, such as metals, ceramics, and polymers, a notable interest in tribological studies in dentistry has also developed [1]. The study of the fundamental processes of lubrication, wear, and friction of opposing dental surfaces that are involved in relative motion and contact within the oral environment is known as dental tribology. The large number of research investigations on the subject concerning human dental tribology is not surprising. Dental implants are a popular option when a natural tooth is badly damaged and replacement is required. They are utilized when a natural tooth is missing due to either periodontal disease or cavities [2]. It is imperative to get trustworthy and insightful knowledge about oral lubrication due to the growing emphasis on oral health and the urgent need for individualized oral care products. The tremendous economic benefits that are associated with this area make it even more important. Human saliva is the main lubricant in the mouth cavity, and there has been an increase in interest in oral lubrication from food colloid chemists, biological scientists, and commercial companies interested in the creation of oral care products and medications [3]. The teeth are unquestionably the most significant bio-tribological components and are essential to human existence. Natural or artificial teeth will eventually wear down since they are always being subjected to chemical and mechanical stress [4]. Oral stress is a physiological occurrence that occurs during the lifespan of the teeth. Wearing dental cusps increases the region of contact and occasionally also increases wear. In order to accurately assess the many forms of wear that occur between restorative materials and teeth, a tribological technique is typically used to compare the oral cavity.

One of the primary subfields of bio-tribology is oral tribology, which encompasses all facets of tribology pertaining to oral systems. The oral cavity's primary functions as a human organ are talking and food processing [5].

DOI: 10.1201/9781003717362-7

There will always be considerable tribological activity in the mouth as a result of these oral biomechanical activities. The primary purpose of teeth is to wear them down, either naturally or artificially. This wear is mostly caused by mastication, that is the process of chewing food that encompasses both open and closed phases. Teeth wear is also caused by thegosis and bruxism, in addition to mastication. These days, chemical factors—primarily from consuming a lot of acidic drinks—play a bigger and bigger part in tooth deterioration. These days, tooth flexure is frequently cited as a potential reason for non-carious cervical tooth lesions. Atypical oral wear, lubrication, and friction are linked to a number of oral ailments. Tooth wear is a medical condition that is getting worse. Significantly, among aging populations, knowing oral tribology may aid in medical prevention and treatment by illuminating the elements that lead to oral illnesses. Comprehending the reasons behind the exceptional tribological behavior of human teeth will facilitate the development of future wear-resistant engineering solutions that offer longer service lives by mitigating wear and friction.

7.2 DENTAL ORGAN: COMPONENTS, ATTRIBUTES, AND CONNECTIONS

The study of wear, lubrication, and friction among the oral cavity's components is known as dental organ tribology. It focuses on the external surfaces of the tooth and dental parts during different oral activities, including speaking and chewing. Greek terms "tribos," which means rubbing or sliding, and "logos," which means study, are combined to form the word "tribology." Tooth enamel is an essential component that plays a crucial role in maintaining the durability and dependability of the entire stomatognathic system, not just the dental organ. Research on the special qualities of enamel is therefore crucial. The research of enamel's properties faces methodological challenges due to its variety, primarily related to the observation scale. The dental organ ought to be viewed as a system that needs lubrication to function. Saliva is found in the material plane, according to Czicho's technique, and its primary purpose is to lubricate the teeth in contact with one another and with the soft tissues of the mouth. Saliva's excellent lubricating qualities lessen wear among teeth, among teeth and food, and among teeth and the mucosa or tongue [6, 7]. The neuromotor system linked to the operation of the dental organ recognizes the variety of food that goes through its early development in the mouth cavity. Appropriate load levels and kinematic characteristics must be included in the system input to guarantee the efficacy of chewing process. The unique dynamics that occur throughout a single chewing cycle define the input parameters. The forces applied to the teeth when biting and chewing vary greatly and are influenced by the kind of food being eaten as well as the tooth's location within the dental arch.

Understanding the interaction of force and pressure is more crucial from a tribological perspective than understanding the normal forces. Determining the pressure during mastication, however, presents serious methodological challenges [8, 9].

Probably, the parameter with the most dynamics of all those that could influence the tribological processes involving the tooth organ is the environmental chemistry. Alterations are inherently linked to the consumption of a range of foods, although they can also result from specific illnesses. The high level of acidity in the environment is particularly significant in this case since it not only has powerful chemical effects but also has the potential to exacerbate tribological processes, which negatively impact enamel durability. However, it is more practical to include widely recognized tribological characteristics, such as the friction coefficient and/or wear intensity, in addition to when analyzing a dental organ as a tribological system. Tribological analyses of the tooth organ raise a number of questions about how natural materials behave when subjected to friction. The question is entirely justified because there are many different study approaches and methodologies that are used in practice, which frequently result in contradictory findings about the same topics. The field of biomedical engineering has experienced significant advancements in bio-tribology-related fields. Notable advancements have been made in the use of novel tools, materials, and technologies, as well as in the elucidation of processes related to the friction phenomena in living things. Bio-tribologists are also interested in dentistry since it deals with processes and phenomena associated with the stomatognathic structure and its principal component, the dental organ [10]. Recognizing the wear patterns and frictional forces that arise when molars, premolars, and molars come into contact with one another, as well as when biting and grinding are performed. Examining how dental materials—such as prosthesis, restorative materials, and orthodontic appliances—behave tribologically when they are brought into interaction with natural teeth in the course of daily activities, and studying the lubricating qualities of oral fluids, such as saliva, which are essential for minimizing friction between tooth surfaces during chewing motions. Decreasing wear and damage by researching the implications of chewing forces, crushing, and other mechanical stressors on the outer layers of teeth and dental materials, and examining the tribological features of the temporomandibular joint, which are important in jaw movement when speaking and chewing.

Materials for dentistry that are more wear-resistant offer greater lubrication in the environment of the mouth, and improve the general functionality of restorations for teeth and prosthetics, which are developed as a result of investigation in dental organ tribology. By studying the tribological elements of the dentistry organ system, dental practitioners are able to develop and execute improved treatments and preventive initiatives for preserving oral health. Principles from dentistry, biomechanics, and materials science are all applied in this multidisciplinary field.

7.3 COMPREHENDING ORAL TRIBOLOGY

Knowing the fundamentals of tribology and how they relate to the interactions in the oral cavity is necessary to comprehend oral tribology. When it comes to the oral cavity, oral tribology focuses on the wear, lubrication, and friction that happen between different oral structures during activities, such as speaking, chewing, and swallowing. Researching the tribological behavior of dental surfaces as you move, comprising the enamel, the dentin, and the cementum. This entails investigating the tribological characteristics of dental materials, including restorations such as bridges, crowns, and orthodontic appliances, as well as the manner in which such surfaces react during chewing, crushing, and other oral actions. This involves taking into account how these materials react to stresses from movement in the oral environment and interact with natural teeth.

Taking into account the masticatory muscles' function in oral tribology, wherein muscular contractions are crucial in producing forces during chewing, and understanding the mechanical aspects of mouth functions and their impact on interacting surfaces by incorporating biomechanical ideas into the study of oral tribology and utilizing oral tribology findings in clinical dentistry in order to enhance prostheses, dental materials, and treatment plans. Creating materials resistant to the tribological difficulties of the mouth environment is one aspect of this. The advancement of dentistry's supplies and treatment modalities, which seek to maximize the effectiveness and longevity of dental procedures while reducing wear and injury to oral tissues, depends on research in oral tribology. To handle the intricate interactions within the oral cavity, dental experts, biomaterials scientists, and engineers must work together.

Numerous biological systems are studied in bio-tribology, and significant discoveries made in one system may frequently be applied to another with equal significance. The identification of distinct tribological mechanisms in biological systems should offer significant new perspectives for the overall development of tribology. Tribology method and technique advancements are equally crucial to the advancement of bio-tribology research. All facets of tribology pertaining to oral systems are included in oral tribology. The mouth cavity in humans is made up of the tongue, palate, chin, teeth, mucosa, and glands. Studies on tooth structure, saliva, and the soft tissues of the oral cavity are all part of oral tribology. To replicate the tribological behavior of the oral system, four types of test methodologies have been employed: finite element analysis, in vitro examination, in situ examination, and clinical examination in vivo. Enamel, dentine, pulp, and dentine–enamel junction make up human teeth. A developing trend is the increasing number of tribological studies pertaining to an increasing number of biological systems. Because of regular oral function, dental wear and friction are unavoidable lifetime processes. On the other hand, human teeth are the body's most resilient organ [11].

In dentistry, metallic biomaterials are commonly employed to create orthodontic devices, endodontic instruments, maxillofacial surgical structures, frameworks, and dental implants. The primary issue surrounding these instruments is the damage caused by tribo-corrosion that metals encounter in the mouth cavity. The included studies examined the characteristics of cpTi and alloys, including TiAlV and TiZr, as well as implants and framework applications. Dental implants' material behavior is greatly influenced by their surface features. To improve their wear and corrosion resistance, numerous surface changes have been made. Because of its exceptional mechanical qualities, great biocompatibility, and resistance to corrosion, titanium and its related alloys are among the most often utilized materials in dental applications. The oral environment for metallic implants is extraordinarily complicated and highly aggressive when seen from a tribological perspective. The presence of saliva and germs can induce multiple implications for the bio-tribo-corrosion of dental implants. Material degradation can promote crestal resorption of bone and further increase the possibilities for a larger intensity of micro-movement and failure of implants. The techniques utilized to decontaminate implant surfaces in the oral cavity, such as simultaneous, chemical, and mechanical processes, may also contribute to an increase in surface tribo-corrosion. Osteoblasts are able to ingest and phagocyte wear debris throughout this natural defense process, changing the fundamental cell mechanisms and causing necrosis in the cell. Despite recent advancements in science, research on the full implications of metal degradation products in the underlying causes of peri-implantitis remains inconclusive [12, 13].

Future research is needed to adapt the oral ecological condition's complexity to the bench, taking into account every important element related to the bio-tribo-corrosion phenomenon. The recognition comprehending the pertinent variables may enable the creation of a streamlined tribo-corrosion test that might include every element required for experimental evaluations. Moreover, the development of in vivo tribo-corrosion analysis is required for a clinically relevant trial test. The oral environment, which includes cyclic stresses, saliva, periodontal fluids, oral biofilm, and micro-movements, presents a difficult scenario for the implant material. Therefore, it is crucial to comprehend the tribo-corrosion pushing mechanisms in such circumstances.

7.4 TRIBOLOGICAL STUDIES IN DENTISTRY AND TOOTH WEAR

Tribology is the science and engineering of interacting surfaces in relative motion, and it plays a significant role in dentistry, particularly in understanding tooth wear. Tooth wear is a complex process influenced by various mechanical, chemical, and biological factors. Tribological studies in dentistry focus on investigating the mechanisms, causes, and prevention of tooth wear. The contact of teeth with abrasive particles found in food and toothpaste causes abrasive wear. Chemical reactions that can disintegrate and soften the tooth structure, such as contact with acidic substances, cause erosive wear. Grinding and chewing are two common oral functions that result in attritional damage. The adhesion of surfaces that come into contact during mastication, can happen in tooth wear between opposing tooth surfaces. The rubbing effect of abrasive particles wears down surfaces, and regular physiologically tooth-to-tooth contact during chewing causes a gradual deterioration of tooth structure. Tribological investigations use a range of diagnostic methods, including atomic force microscopy (AFM), profilometry, and scanning electron microscopy (SEM), to evaluate tooth wear [14–20]. These methods aid in the visualization and measurement of tooth surface wear, and it is essential to comprehend the tribological features of dental wear in order to devise preventive strategies. This involves encouraging the use of dental implants with improved wear resistance, dietary changes, and appropriate oral hygiene practices.

The creation of dental implants and the coatings that can tolerate wear and friction is aided by tribological investigations. The goal of ceramic material advancements for dental restorations is to enhance their tribological characteristics. Tooth wear can be influenced by individual patient characteristics, including malocclusion, teeth grinding, and systemic diseases. Understanding the interaction between these elements and wear patterns is aided by anthropological research [21]. In order to obtain a thorough understanding of tooth wear, tribological research in dentistry frequently takes a multidisciplinary strategy, combining knowledge in the science of materials, engineering, and dentistry [22]. Tribological studies are essential to the advancement of dental wear research. This information aids in the creation of preventive methods, better dental materials, and better patient care. To quantify the degree of dental wear and identify the different forms of wear among teeth and restorative materials, the oral cavity is often contrasted to a tribological system. Numerous studies have concentrated on enhancing the wear characteristics of dental materials and shielding teeth from excessive wear because dentists have the most influence on the selection of materials. Dental tribology research has received a lot of attention since the discovery of biomaterials. Many wear simulation tools have been created. Most people compare the oral cavity to the tribological system. Food pellet fragments, saliva, air, teeth, and opposing teeth make up the system. Although tooth wear has been extensively studied, we still do not fully grasp its etiology and pathology.

Various dental materials were employed in the experimental investigations, and different authors created distinct tribometer testing techniques for measuring tooth wear. The amount and depth of wear at the oral contact areas are the main factors influencing quantitative assessment, whereas comprehensive topographic surface analysis is the basis for qualitative assessment. Consequently, a variety of approaches and morphological and microscopic procedures have been employed to evaluate the loss or destruction of dental components and teeth both in vivo and in vitro. Tooth wear is caused by a variety of variables that interact mechanically, chemically, and biologically. Tooth deterioration should be promptly detected and treated as a clinical concern. To replicate the tooth wear environment, a number of tools and techniques have been developed in earlier research; still, the procedure is rather intricate.

7.5 METHOD OF TOOTH WEAR ASSESSMENT

Assessing tooth wear entails determining the degree and origins of erosion on the teeth. Dental practitioners utilize a variety of techniques to evaluate tooth wear. Several of these are that dentists visually inspect teeth to find wear indicators. This can entail keeping an eye out for modifications to the color, texture, and shape of the teeth. It is possible to snap pictures to record the degree of tooth wear. This gives the patient and the dentist a visual record and aids in tracking changes over time. Research models, or copies of a patient's teeth, can be made using dental impressions. With the aid of these models, dentists may plan treatments and evaluate tooth deterioration in three dimensions. Different dental wear indexes have been created in order to measure and standardize the amount of wear. Systems for classifying and rating tooth wear include the Eccles Index and the Smith and Knight Dental Wear Index. Information about tooth wear can be captured in close-up photographs. These pictures can be utilized for record-keeping and to inform patients on the state of their teeth. Using a micro-hardness tester, surface micro-hardness testing determines how hard the surfaces of teeth are. It offers details on the dental tissues' resistance to abrasive forces. To capture tooth surface details, impressions made of polyvinyl siloxane or alginate might be utilized. Replicas can then be made using these imprints to look more closely at wear patterns.

Dentists might employ articulating paper to highlight contact locations between teeth. Analyzing the paper markings can help identify early contacts and abnormal wear patterns. This entails employing scanning electron microscopy to analyze dental micro-wear patterns. To better understand tooth wear, micro-wear analysis might shed light on eating and chewing patterns. Teeth wear can be evaluated in three dimensions using cone-beam computerized tomography or similar 3D imaging methods. This facilitates the comprehension of wear severity and spatial distribution. Knowing a patient's habits, nutritional needs, and way of life can help determine whether bruxism, acidic erosion from diet, or occupational variables are the probable reasons for tooth wear. When multiple techniques are combined, tooth wear can be thoroughly assessed, assisting dentists in identifying underlying causes and creating effective treatment strategies.

To provide a more exact simulation of the teeth sliding movement, more sophisticated reciprocating wear test devices have been used. In order to simulate physiological movement, a test rig for abrasion and attrition research has been developed. It extends the concept of unidirectional sliding by loading and driving an enamel-tipped stylus across a test material in a single direction while an abrasive slurry is present. Like any test simulation, the rigs have a number of flaws that limit how accurately they simulate the real world. These flaws could include the use of flat specimens, the use of water as a lubricant, or the absence of an abrasive slurry. Similar to test rigs used to examine wear from tooth contact, artificially produced tooth brushing test rigs come in a variety of designs and provide more control over experimental parameters than in vivo testing. However, they share the same drawbacks as their contact wear equivalents. In situ examination offers an incomplete balance between in vivo and in vitro. Terms related to wear mechanisms in clinical dentistry are not the same as those used in traditional engineering tribology. The mechanisms contributing to tooth wear have been analyzed, and it is clear that the majority of wear issues are related to occlusal contact. The noninvasive wear mechanisms, such as open-stage chewing and brushing their teeth, have a reduced effect on enamel wear.

The running-in process that causes an occlusal contacting wear mechanism is common to wear in engineered component contacts. As a result of decreased wear rates and more conformal contacts, this may actually be advantageous. Wear during tooth brushing owing to abrasion particles in the toothbrush appears to be solely a concern with the dentine and some materials for restoration. The acidic nature of the oral environment has a considerable impact on all the various wear mechanisms. Tooth wear studies employ a wide range of test platforms. It has been extremely difficult to relate findings to actual in vivo tooth wear in the majority of cases. The fact that there are no reliable methods for measuring attrition in living teeth further exacerbates this.

7.6 TRIBOLOGICAL EXAMINATION

Various operating methods have been devised by various researchers by employing tribometers to perform wear testing, and various materials for dentistry have been employed in the experimental investigations. The scientific literature's experimental analyses vary depending on the supplies, techniques, and equipment employed. Moreover, the mechanical characteristics, corrosion, and wear resistance of restorative materials are occasionally evaluated in comparison to other materials, such as metal alloys. The natural tooth paired with dental ceramics—the most popular material for dental restorations—is the most typical arrangement. Two geometries are required to do the tests: a cylindrical pin and a flat sample. This is because the tribometers often replicate the contact in pins-on-flat or ball-on-flat arrangements. In most cases, a dental cusp is used to test a natural tooth on dental ceramics, and a flat ceramic sample is created. However, when a contact tooth is replicated toward the tooth, both the pins and focuses on are made from natural teeth, and because the surface geometry of natural teeth is irregular, it is difficult to control the real contact surface. In dental tribology, three types of tribometers are commonly used: reciprocating tribometers, ball-craters, and chewing simulators.

The oral cavity and teeth are examined medically during a dental examination. Tribological concerns become significant when analyzing the resistance to wear of dental restorations such as bridges, crowns, and implants. Designing dental restorations that are strong and long-lasting requires an understanding of how these substances react in the oral environment. Biomechanical and tribological aspects play a role in the interaction between dental implants and surrounding tissues. The long-term viability of dental implants can be influenced by appropriate design and material selection. Friction and wear are the factors that are taken into account while studying the relationship between teeth during biting and chewing. Anomalies, including malocclusion or misalignment, can cause unusual tooth wear patterns. For the sake of patient comfort and prosthetic device longevity, the wear and friction characteristics of the materials that are utilized for prosthetic teeth, dentures with partials, and other prosthetic teeth are crucial. To ensure that dental materials can endure the mechanical forces and wear involved with routine oral functions, efforts are being made to further develop the tribological characteristics of dental materials in materials science and dentistry research.

7.7 POSSIBLE PROGRESS IN ORAL TRIBOLOGY

Understanding how materials interact in dental applications—particularly when it comes to wear, friction, and lubrication—is made possible in large part by the study of tribology. Dental technology and materials have advanced due to the field of tribology, which has improved patient outcomes, longevity, and performance. When designing and choosing materials for dental implants, anthropological factors play a crucial role. In order to improve the tribological properties, lower wear, and increase the longevity of the implant-tissue contact, coatings on the surface and modifications have been advanced. It is essential to comprehend the mechanisms of wear at the interface between an implant and an abutment and how they interact with the surrounding bone. The creation of new materials with enhanced tribological properties could be an advancement. Tribology contributes to the enhanced functionality of dental prosthetic materials, including bridges, crowns, and dentures. Materials that can resist the mechanical strains brought on by chewing and biting while minimizing wear are the focus of research. In dental prosthesis, wear mechanisms occur when various materials, including ceramics and natural teeth, come into contact with one another. Utilizing cutting-edge ceramics and coatings to lower wear rates is one example of an advancement. It is essential to comprehend the tribological features of tooth occlusal contact in order to prevent aberrant wear patterns. Improved malocclusion and misalignment diagnosis and treatment options could be among the advancements. The analysis of contact stress and the forces of friction between opposing teeth is a key component of tribological considerations in occlusion. The creation of materials that can endure these forces without experiencing undue wear and tear is one possible advancement.

Enhancing the resistance to wear of dental restoration materials, such as composites and amalgams, is the main goal of tribological research. This is especially crucial in oral regions that experience a lot of stress. Food particle abrasion, chemical wear, and interactions with neighboring teeth are the wear processes in restorative materials. The addition of metal nanomaterials or additional ingredients to improve wear resistance could be considered an advancement. When developing dental care goods, such as toothpaste and toothbrushes, anthropology is important. Potential advancements could include the incorporation of abrasives and maximizing the effectiveness of brushing bristle materials to achieve efficient cleaning without unduly wearing down enamel. Tribological factors come into play when brushing bristles, where saliva particulates and tooth surfaces interact. Improvements could concentrate on reducing enamel wear and increasing cleaning effectiveness. Tribology is valuable to dentistry because it helps create dental materials and technology that are resilient to the intricate mechanical and chemical environments found in the oral cavity. The goal of ongoing research is to enhance the tribological characteristics of dental materials, which will increase their endurance and improve oral health results in general.

These developments will probably be driven by interdisciplinary cooperation among materials scientists, biologists, engineers, and dentists as oral tribology research progresses. These possible areas of advancement could increase patient comfort, lengthen the lifespan of dental materials, and boost the effectiveness of numerous dental procedures as a whole.

7.8 CONCLUDING REMARKS

Within the context of bio-tribology, there is a growing relationship between tribology and dentistry. According to the results of a bibliographic search, the number of articles on this subject has increased over the past ten years from a few dozen to over 3000, and this tendency is continuing. To improve our knowledge of dental tribology and the materials used in dental restorations, a great deal of research is being done right now. In our view, one noteworthy aspect to take into account is the relatively low usage of the terms "tribology" and "bio-tribology" within the medical domain. Bibliographical research would certainly benefit from their increased use, which would better define this field of study.

The materials chosen for dental restorations are undoubtedly a very important factor. A few fundamental qualities that these need to possess include biocompatibility, non-cytotoxicity, corrosion resistance, and wear resistance. Composites, ceramics, metals, and alloys have all been employed extensively for dental restorations, although these materials have different issues than real tooth enamel. Certain materials exhibit low biocompatibility, and in certain instances, it has been demonstrated that they are cytotoxic. Dental composites typically exhibit insufficient wear resistance; in contrast, ceramics respond better to the oral environment.

Stress the need to know oral tribology for those working in the biomaterials, prosthodontic, and dental areas. The oral environment presents particular difficulties, and knowledge from tribological research might enhance the performance and lifetime of dental materials. The complicated interplay of various surfaces in the oral cavity, including natural ones, restorations for dentistry, and prosthetic devices, was discussed and emphasized how important tribological factors are to the efficacy of these interactions, preserving oral health, and saliva's function as a lubricant, and the ways in which its content and flow affect tribological behavior in the oral environment are acknowledged. Comprehending the variables is essential for creating materials that can perform well when saliva is present. The impact of different dental materials on wear, lubrication, and friction is taken into consideration while designing prostheses and choosing materials to improve their functionality.

Promoting cooperation between material scientists, dentists, tribologists, and other pertinent professions. Because of its complexity, oral tribology needs a multidisciplinary approach to fully handle its many issues. In summary, the dynamic area of oral tribology is essential to the improvement of dental procedures and prosthetic devices. As we gain more knowledge, new opportunities for innovation arise, improving patient well-being and the state of oral health.

REFERENCES

1. Zhou, Z.R., Jin, Z.M., Biotribology: Recent progresses and future perspectives. *Biosurface and Biotribology*, 1, 3–24, 2015.
2. Zheng, Yong, Bashandeh, Kian, Shakil, Ahmad, Jha, Swarn, Polycarpou, Andreas A., Review of dental tribology: Current status and challenges. *Tribology International*, 166, 2022.
3. Shellis, R.P., Addy, M., The interactions between attrition, abrasion and erosion in tooth wear. *Monographs in Oral Science*, 25, 32–45, 2014.
4. Balasubramaniam, R., Paesani, D., Koyano, K., Tsukiyama, Y. Sleep bruxism: A comprehensive approach to clinical practice. *Contemporary Oral Medicine*, 2267–2301, 2019.
5. Powers, J.M., Bayne, S., Friction and wear of dental materials, in *Handbook of Friction and Wear* (Materials Park, ASTM, 1988), pp. 666–678.

6. Sarkar, A., Andablo-Reyes, E., Bryant, M., Dowson, D., Neville, A., Lubrication of soft oral surfaces, *Current Opinion in Colloid & Interface Science*, 39, 61–75, 2019.
7. Dodds, M.W.J., Johnson, D.A., Yeh, C.-K., Health benefits of saliva: A review, *Journal of Dentistry* 2005, 33, 223–233.
8. Laurell L., Occlusal forces and chewing ability in dentitions with cross arch bridges, Swed. *Dental Journal*, Suppl. 26, 1–45, 1985.
9. Kim, K.-S., Choi, J.-H., Kim, S.-T., Kim, C.-Y., Ahn, H.-J., Bite force, occlusal contact area and occlusal pressure of patients with temporomandibular joint internal derangement, *Korean Journal of Oral Medicine*, 31, 265–273, 2006.
10. Sajewicz, E., A comparative study of tribological behaviour of dental composites and tooth enamel: An energy approach, *Proceedings of the Institution of Mechanical Engineers, Part J: Journal of Engineering Tribology*, 224, 559–568, 2010.
11. Bachmann, T., Klän, S., Baumgartner, W. et al., Morphometric characterisation of wing feathers of the barn owl Tyto alba pratincola and the pigeon Columba livia. *Frontiers in Zoology*, 4, 23, 2007.
12. Alves, A.C., Oliveira, S., Wenger, F., Ponthiaux, P., Celis, J.P., Rocha, L.A., Tribocorrosion behaviour of anodic treated titanium surfaces intended for dental implants. *Journal of Physics D: Applied Physics*, 46, 404001, 2013.
13. Alves, S.A., Rossi, A.L., Ribeiro, A.R., Toptan, F., Pinto, A.M., Shokuhfar, T., Improved tribocorrosion performance of bio-functionalized TiO(2) nanotubes under two-cycle sliding actions in artificial saliva. *Journal of the Mechanical Behavior of Biomedical Materials*, 80, 143–154, 2018.
14. Wetselaar, P., Wetselaar-Glas, M.J., Katzer, L.D., Ahlers, M.O., Diagnosing tooth wear, a new taxonomy based on the revised version of the Tooth Wear Evaluation System (TWES 2.0), *Journal of Oral Rehabilitation*, 47(6), 703–712, 2020.
15. Lanza, A., Ruggiero, A., Sbordone, L., Tribology and dentistry: A commentary. *Lubricants*, 7(6), 52, 2019.
16. Rees, J.S., Somi, S., A guide to the clinical management of attrition. *British Dental Journal*, 224(5), 319–323, 2018.
17. De Baat, C., Verhoeff, M., Ahlberg, J., et al., Medications and addictive substances potentially inducing or attenuating sleep bruxism and/or awake bruxism. *Journal of Oral Rehabilitation*, 48(3), 343–354, 2021.
18. Khayat, N., Winocur, E., Kedem, R., Winocur, Arias O., Zaghal, A., Shpack, N., The prevalence of temporomandibular disorders and dental attrition levels in patients with posterior crossbite and/or deep bite: A preliminary prospective study. *Pain Research & Management*, 1, 8827895, 2021.
19. Badavannavar, A.N., Ajari, S., Nayak, K.U.S., Khijmatgar, S., Abfraction: Etiopathogenesis, clinical aspect, and diagnostic-treatment modalities: A review, *Indian Journal of Dental Research*, 31(2), 305–311, 2020.
20. Sato, T., Fukuzawa, Y., Kawakami, S., et al., The onset of dental erosion caused by food and drinks and the preventive effect of alkaline ionized water. *Nutrients*, 13(10), 3440, 2021.
21. Bammidi, R., Prasad, K.S., Nagaraju, B., Synthesis and characterization studies of Ag-Nps for endodontics, in *Recent Advances in Material Sciences* (Singapore, Springer, 2019), pp. 541–547.
22. Bammidi, Roopsandeep, Prasad, Siva, Ti-6AL-4V as dental implant, *EAS Journal of Dentistry and Oral Medicine*, 2(1), 2020.

Chapter 8

Performance estimation of circular bearing with nano-lubricants

Sandeep Soni

8.1 INTRODUCTION

The hydrodynamic lubrication theory, which was developed from a fundamental fluid mechanics equation with a few constrictive assumptions, is governed by the Reynolds equation [1].

An analytical method was used to reduce the Reynolds equation to algebraic form, and static characteristics were obtained by adopting a finite element numerical procedure. The idea of an infinitely long journal bearing serves as the foundation for many of the practical fixes for oil-lubricated circular journal bearings [2].

Nanotechnology has the potential to completely transform every aspect of contemporary life. It is having a significant impact on every industry, including manufacturing, education, healthcare, and defense. In the past 20 years, tribology has benefited from the application of nanotechnology, particularly in the areas of thin-film lubrication and nanoparticle lubricant additives [3, 4]. When added to the base oil as lubricant additives, nanoparticles significantly improve the tribological characteristics of the base oil. Since energy shortage will likely be a common occurrence in the future, friction and wear losses must be minimized [5, 6].

Lubricants are employed in support mechanisms, such as journal bearings, which are widely utilized in high-load and high-speed applications. These bearings suffer from frictional loss and surface wear. These days, hydrodynamic journal bearings use nano-lubricants, and the utilization of these materials is expanding [7, 8]. Numerous tribologists have been concentrating on the impact of nano-lubricants in hydrodynamic journal bearings in the past few years. Friction and wear losses on contact surfaces are generally minimal at low loading. However, the lubricant film formed between the journal and the bearing will be severely compressed at high loads, and it might not be able to shield contact surfaces [9].

According to certain studies, adding nanoparticles to a mixture strengthens the supporting forces [10]. Therefore, if nanoparticles are added to the lubricant, wear and friction can be decreased. Over the past 10 years, a lot of research has been done to examine the characteristics of nano-lubricants and

DOI: 10.1201/9781003717362-8

their uses. Numerous kinds of nanoparticles have been employed as lubricant additives, and research has been done on how well they work in thin-film lubrication [11]. The application of nanoparticle lubricant additives reduced wear and friction in tribo-surfaces, according to all of these experiments. Regarding the theoretical simulation of nano-lubricant in fluid-film bearing, there is a significant research gap. Existing research indicates that by taking into account solely the variable viscosity analysis of the nanoparticle additive, performance attributes, both static and dynamic, can be improved [12].

According to [13] the investigation on the function of nanoparticles in nano-oil lubrication, adding nanoparticles may increase the lubricant's viscosity. A small number of studies have also shown that employing nanoparticles as lubricant additives, which are distinguished by thick oil layers separating the tribo-surfaces, has good effects [14, 15].

A handful of analytical studies [16, 17] have demonstrated an enhancement in the stability of hydrodynamic bearings operating on aforesaid nano-oils. It is still not possible to create a generalized theory that would accurately model the performance characteristics of journal bearings that use nanoparticle additives. An analytical model that takes additive concentrations into account has been developed by [18, 19] for simulating the static properties of journal bearings running on engine oil containing TiO_2, Al_2O_3, and ZnO nanoparticles. The theoretical analysis of journal bearing approximations has been well documented in [20]. The review of the nano-particles effect on the tribological effectiveness of oil lubrication has been described in [21]. The work [22] proposes a straightforward yet exhaustive model to foresee the effective viscosity of nano-fluids within this framework. The Krieger–Dougherty hypothesis for effective viscosity of liquid-solid suspensions is the foundation of the current model, which is constructed by including the effects of the interfacial layer and aggregation.

This chapter presents the impact of TiO_2, WS_2, and CuO nanoparticle concentrations in the particular base fluid (SAE 30) on the static performance parameters of a fluid-film bearing.

8.2 GOVERNING EQUATIONS

This section covers the implementation of a viscosity model for nano-lubrication modeling. Furthermore, the modified Reynolds equation with nanoparticle variations has been developed to compute the static performance parameters for journal bearing analysis.

8.2.1 Nano-lubrication modeling

Several mathematical models have been established over the years to replicate particle suspension viscosities. Inter alia, modified Krieger–Dougherty model is contemplated to simulate fluid viscosities, which are in good compliance with nano-fluid experimental viscosities [4, 8]. The validity of the modified Krieger–Dougherty model in resembling the shear viscosities of nanoparticle (TiO_2, WS_2, and CuO) dispersion in base oil (SAE 30) is inquired. The Krieger–Dougherty model is manifested as:

$$\mu_{\text{NanoFluid}} = \mu_{\text{BaseFluid}} \left(1 - \frac{\xi}{\xi_m}\right)^{-\{\kappa\}*\xi_m} \tag{8.1}$$

where ξ is the suspension's particle concentration (volume fraction). ξ_m is the maximum particle packing fraction in the suspension, $\{\kappa\}$ is known as inherent viscosity, and it is usually given a value of 2.50 for the single dispersion of spherical particles [4, 8]. The original Krieger–Dougherty model is revised by [4, 17, 18], including the packing fraction within the nanoparticle aggregates. The modified Krieger–Dougherty model is manifested as:

$$\mu_{\text{NanoFluid}} = \mu_{\text{BaseFluid}} \left(1 - \frac{\xi_a}{\xi_m}\right)^{-\{\kappa\}*\xi_m} \tag{8.2}$$

where ξ_m has typically a value of 0.50 and 0.605 for low-shear and high-shear applications, respectively. In this work, $\xi_m = 0.605$ has been considered for the journal bearings. ξ_m is the effective aggregate volume fraction calculated as follows:

$$\xi_a = \left(\frac{\xi_p}{\xi_{ma}}\right) \tag{8.3}$$

where ξ_p is the particle volume packing fraction and ξ_{ma} denotes the maximum packing fraction within individual aggregates. A power law approach for calculating aggregate volume fraction has been presented by [4, 18] and can be customized as follows:

$$\xi_a = \xi_p * \left(\frac{\delta_a}{\delta}\right)^{(3-D)} \tag{8.4}$$

D is a measure of the variance in packing percentage inside each aggregate nanoparticle in the suspension and is referred to as the fractal index. The value of D depends on the particle aggregation mechanism.

Values of D range from 1.60 to 1.80 for aggregates formed by diffusion and from 1.80 to 2.30 [3, 17] for aggregates formed by constituent reactions. In general, 1.80 is regarded as the benchmark value for nano-fluids based on TiO_2 particles [4, 18]. δ_a and δ denote the radii of the aggregate particle size and primary particle size, respectively. Finally, the modified Krieger–Dougherty viscosity model for nano-lubricant is demonstrated as [17]:

$$\mu_{\text{NanoFluid}} = \mu_{\text{BaseFluid}}\left(1-\frac{\xi_a}{\xi_m}\right)^{-\{2.50\}*\xi_m}$$
$$\xi_a = \xi_p * \left(\frac{\delta_a}{\delta}\right)^{(3-D)} \tag{8.5}$$

Equation (8.5) has been used to modify the Reynolds equation for predicting the static performance of a long-approximated bearing utilizing nano-lubricants containing TiO_2, WS_2, and CuO nanoparticles with the base fluid (SAE 30).

8.2.2 Modeling of Reynolds equation

Taking into account the dimensional Reynolds equation provided by [23] for laminar flow in an iso-viscous Newtonian lubricant, it can be stated as follows:

$$\frac{\partial}{\partial x}\left(\frac{h^3}{\mu}\frac{\partial p}{\partial x}\right)+\frac{\partial}{\partial z}\left(\frac{h^3}{\mu}\frac{\partial p}{\partial z}\right)=6U\frac{\partial h}{\partial x} \tag{8.6}$$

Since $\partial p / \partial z \cong 0$, for a long bearing, the second term on the LHS of Equation (8.6) can be eliminated, and the Reynolds equation in 1D becomes the following [23]:

$$\frac{\partial}{\partial x}\left(\frac{h^3}{\mu}\frac{\partial p}{\partial x}\right)=6U\frac{\partial h}{\partial x} \tag{8.7}$$

The Reynolds equation in its non-dimensional form is as follows [23]:

$$\frac{\partial}{\partial \theta}\left(\bar{H}^3\frac{\partial \bar{p}}{\partial \theta}\right)=6\bar{\mu}\frac{\partial \bar{H}}{\partial \theta} \tag{8.8}$$

The following non-dimensional terms are used to build the non-dimensional form of Equation (8.8):

$$x=(\theta * R),\, h=(c*\bar{H}),\, \bar{\mu}=(\mu_{\text{NanoFluid}} / \mu_{\text{BaseFluid}}),\, p=\left(\frac{\mu_{\text{BaseFluid}} * U * R}{c^2}\right)\bar{p}$$

8.3 FINITE ELEMENT ANALYSIS

The Reynolds equation (Equation 8.8) has been solved by utilizing the finite element technique. A computer algorithm is used to conduct numerical research investigations to determine various bearing parameters. A four-noded element has been used to discretize the lubricant regime, and the pressure in the element is obtained as follows [2]:

$$\bar{p} = \sum_{j=1}^{4} N_j^p \bar{p}_j, \theta = \sum_{j=1}^{n} N_j^p \theta_j$$
$$\frac{\partial \bar{p}}{\partial \theta} = \sum_{j=1}^{n} \frac{\partial N_j^p}{\partial \theta} \bar{p}_j \tag{8.9}$$

where N_j = the shape function of the element, while n = number of nodes in each element. Substituting pressure from Equation (8.9), Equation (8.8) is stated as follows:

$$\frac{\partial}{\partial \theta}\left[\bar{H}^3 \frac{\partial}{\partial \theta}\left(\sum_{j=1}^{4} N_j^p \bar{p}_j\right)\right] - 6\bar{\mu}\frac{\partial \bar{H}}{\partial \theta} = R^e \tag{8.10}$$

where R^e is known as the residue, which is minimized by using Galerkin's technique given by [2]:

$$\iint_{\Omega_e} N_i R^e d\theta = 0 \tag{8.11}$$

By integrating the second-order terms of Equation (8.11) part-wise, the global system of equations in the matrix form is given by [2] as follows:

$$\left[\bar{F}_M\right]^e \{\bar{p}\}^e = \left\{\bar{Q}_V\right\}^e + \Omega\left\{\bar{R}_H\right\}^e \tag{8.12}$$

In Equation (8.12), the various coefficients of matrices for the eth element can be defined as follows:

$$\bar{F}_M^e = \iint_{A\,e} \left\{\frac{\bar{H}^3}{6\bar{\mu}}\left(\frac{\partial N_{Vi}}{\partial \theta}\frac{\partial N_{Vj}}{\partial \theta}\right)\right\} d\theta$$
$$\bar{Q}_V^e = \int_{\Gamma^e} \left\{\frac{\bar{H}^3}{6\bar{\mu}}\left(\frac{\partial \bar{p}}{\partial \theta}\right)\right\} N_{Vi} d\bar{\Gamma}^e$$
$$\bar{R}_H^e = \iint_{A\,e} \left\{\bar{H}^3\left(\frac{\partial N_{Vi}}{\partial \theta}\right)\right\} d\theta$$

The final solution for the pressure distribution in the journal bearing is then accessed by the inverse of Equation (8.12), such that:

$$b = \left[\left\{ \bar{Q}_V \right\}^e + \Omega \left\{ \bar{R}_H \right\}^e \right] \qquad (8.13)$$
$$\left\{ \bar{p} \right\} = b \left[\bar{F}_M \right]^{-1}$$

Hereafter, the iterative procedure of Equation (8.13) provided the circumferential pressure variation at nodal points.

8.3.1 Film thickness and boundary conditions

The following are the expressions for the fluid-film thickness (h) and boundary conditions utilized in this simulation.

$$\begin{aligned}
&h = c * (1 + \varepsilon \cos\theta); \varepsilon = (e / c); \\
&\bar{H} = (1 + \varepsilon \cos\theta); \\
&\bar{p}(\theta, z) = 0 \text{ at } \theta = 0, \theta_r \text{ and at } z = \pm(L / D) \\
&\frac{\partial \bar{p}}{\partial \theta}(\theta, z) = 0 \text{ at } \theta = \theta_r \text{ and at } z = 0 \\
&u = 1 \text{ at } r = R \\
&u = v = w = 0 \text{ at } r = R + h \\
&v = w = 0 \text{ at } r = R
\end{aligned} \qquad (8.14)$$

A solution will exist if these requirements are met; if not, the nodes will migrate from one location to another, and the calculation process will continue until a solution is found. Until the pressure difference at the nodes of the future calculations is <0.10 *TL*, the calculation is performed once again. If the convergence conditions are met, the computation procedure is terminated.

$$\bar{Z}_0 = \left| \frac{\left[\bar{p}^{(m)} \right] - \left[\bar{p}^{(m-1)} \right]}{\left[\bar{p}^{(m-1)} \right]} \right| \times 100 \leq TL \qquad (8.15)$$

8.3.2 Solution procedure

Under Newtonian lubricant flow, the nodal pressures have been figured out and then acquired values have been upgraded with the addition of a nanolubricant with varying volume fraction $(\xi = 1\% \text{to} 5\%)$ of TiO_2, WS_2, and CuO nanoparticles with the base fluid (SAE 30). Meanwhile, the hypothesis of Newtonian fluid lubricants has been applied to nano-lubricants in the estimation of distinct static parameters relating to long journal bearings. In the last stage of discretization, a total of 180 elements are chosen in order to obtain the veracity of the solution for the static parameters of the long journal bearing arrangement under changing volume fraction of specific nanoparticles.

8.4 STATIC PERFORMANCE PARAMETERS

The static parameters of the long bearing $(L/D > 1.0)$ system with nano-lubricant in terms of bearing load capacity (W), bearing attitude angle (φ) , and bearing friction coefficient variable $\left(f\left(R/C\right)\right)$ can be obtained by the following expressions.

8.4.1 Load capacity and attitude angle

The components of load capacity are calculated by the following expressions:

$$W_r = \Sigma W_r^e, \; W_\theta = \Sigma W_\theta^e,$$

where

$$\overline{W}_r^e = -\int_0^{2\pi} (\overline{p})^e \cos\theta d\theta$$

$$\overline{W}_\theta^e = \int_0^{2\pi} (\overline{p})^e \sin\theta d\theta$$

The resultant film reaction and attitude angle are calculated as follows:

$$\overline{W} = \left(\overline{W}_r^2 + \overline{W}_\theta^2\right)^{0.5}$$
$$\varphi = \tan^{-1}\left(\frac{\overline{W}_\theta}{\overline{W}_r}\right) \tag{8.16}$$

8.4.2 Friction coefficient variable

The frictional force on each element is added to determine the frictional force on the journal surface, which is written as:

$$\overline{F}_j = \Sigma \overline{F}_j^e$$

where

$$\bar{F}_j^e = \int_0^{2\pi} \left\{ \frac{4}{\bar{H}} + \frac{\bar{H}_0}{\bar{H}^2} \right\}^e d\theta$$

The following expression is used to calculate the bearing friction coefficient variable:

$$f\left(\frac{R}{C}\right) = \left(\frac{\bar{F}_j}{\bar{\bar{W}}}\right) \tag{8.17}$$

8.5 RESULTS AND DISCUSSION

A computer algorithm has been prepared to accurately figure out the static parameters of journal bearing for L / D of 1.50 under the repercussion of TiO_2, WS_2, and CuO nanoparticle lubricants in the base fluid. In the current numerical results, the volume fraction of nanoparticles of the Krieger–Dougherty model varies from 0.01 to 0.05 in step of 0.01. The value $\xi = 0.0$ designates the base fluid.

Ref. [20] is used to validate the listed results in a graphical format, and the results of the simulations closely match the references that are chosen for the current investigation.

Table 8.1 lists the journal bearing's geometrical and operational parameters that were used in the current numerical analysis.

The static parameters of journal bearing under the consequence of TiO_2, WS_2, and CuO nanoparticle lubricants in the base fluid are illustrated in Figures 8.1–8.7. The influence of the volume fraction of nanoparticles of the Krieger–Dougherty model on the load capacity, the attitude angle, and the friction, coefficient variable is illustrated in Figures 8.3–8.7.

The pressure distribution curve considering the consequence of TiO_2, WS_2, and CuO nanoparticle lubricants in the base fluid is depicted in Figure 8.1. It is observed that the circumferential pressure rises in the journal bearing with the nano-lubricant containing a volume fraction of 0.05 of TiO_2, WS_2, and CuO nanoparticles to the base fluid (SAE 30). This observation is the result of SAE 30's increased viscosity due to the presence of nanoparticles. Higher circumferential pressure is noted in the case of SAE 30 + WS_2 nanoparticles with 0.05 volume fraction at $\varepsilon = 0.70$.

Figure 8.2 shows the variation in maximum pressure ($\bar{P}_{max}$) with eccentricity ratio (ε) of long journal bearings for the base fluid (SAE 30) and TiO_2, WS_2, and CuO nanoparticles at 0.05 volume concentrations in SAE 30. The findings demonstrate the strong agreement found in Ref. [20] and the base fluid (SAE 30) lubricated bearing. With the addition of nanoparticles to the

Table 8.1 Operating and geometrical parameters for the present study

Bearing approximation	*Long bearing*
Length to diameter ratio (L/D)	1.50
Viscosity model	Krieger–Dougherty
Base fluid	SAE 30
Lubricant additives	CuO, TiO_2, and WS_2 nanoparticles
Eccentricity ratio (ε)	0.10–0.90
Volume fraction of nanoparticles (ξ)	0.010–0.050

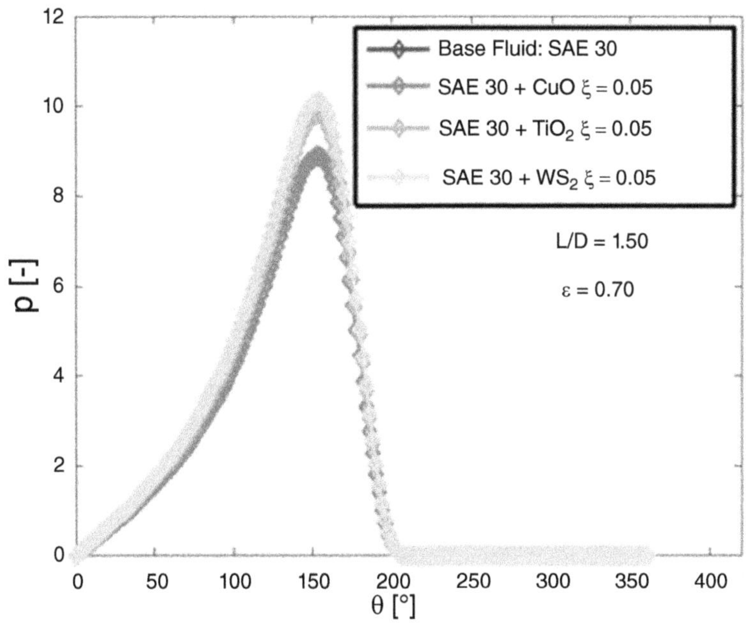

Figure 8.1 Deviation of with for Base Fluid and CuO, TiO_2 and WS_2 nano-lubricants.

base fluid, the maximum pressure also increases. The variation in $\bar{P}_{max}$ for CuO nanoparticles with a 0.05 volume fraction is negligible.

The variation in dimensionless load capacity $(\bar{W})$ with eccentricity ratio (ε) obtained by using TiO_2, WS_2, and CuO nanoparticles at 0.05 volume concentrations in SAE 30 is shown in Figures 8.3 and 8.4. The dimensionless load capacity $(\bar{W})$ is higher due to the presence of WS_2 nanoparticles at 0.05 volume concentrations in SAE 30, as compared to the presence of TiO_2 and CuO nanoparticles at 0.05 volume concentrations in the base fluid, as well as with Ref. [20] and SAE 30 (Figure 8.3a).

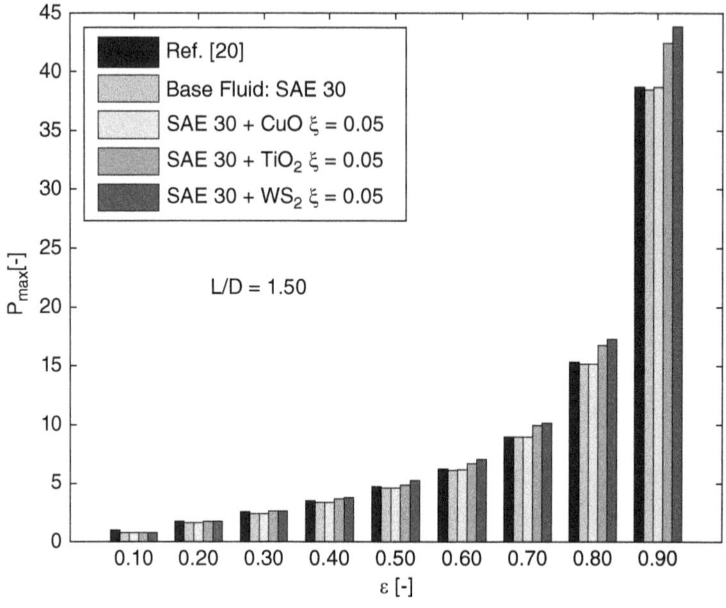

Figure 8.2 Deviation of $\bar{P}_{max}$ with ε for base fluid and CuO, TiO_2 and WS_2 nano-lubricants

It is scrutinized that the dimensionless load capacity ($\bar{W}$) carried by bearings lubricated with base fluid (SAE 30) and the nano-lubricant (CuO, TiO_2, and WS_2 nanoparticles at $\xi = 0.01 \text{to} 0.04$) increases with the increase of eccentricity ratio (Figures 8.3b and 8.4a, b).

Furthermore, it is noteworthy that bearings lubricated with nano-lubricants exhibit higher $\bar{W}$ values than those lubricated with the base fluid. The increase in hydrodynamic pressure is the cause of this rise.

Figure 8.5(a and b) depicts the change in bearing attitude angle (ϕ) with eccentricity ratio (ε) obtained across CuO and WS_2 nanoparticles at 0.01–0.04 volume concentrations in SAE 30. The graphic makes it evident that, for nano-lubricants, the attitude angle falls as the eccentricity ratio increases. It suggests that the volume percentage of nanoparticles up to 0.04 in the base lubricant (CuO and WS_2 nanoparticles) less alters ϕ.

The influence of the addition of TiO_2 nanoparticles in the base fluid (SAE 30) brings curtailment in the attitude angle (ϕ) with an increment of eccentricity ratio (ε) from 0.10 to 0.90 (Figure 8.6). The shaft position and the lubricant film's wedge effect both improve with an increase in the volume concentration of TiO_2 nanoparticles from 0.01 to 0.04 in SAE 30.

Figure 8.7 illustrates the deviation of the friction coefficient variable ($f\left(R/C\right)$) with distinct eccentricity ratio (ε) of a journal bearing operating with a base fluid and nano-lubricant. As depicted in the figure, $f\left(R/C\right)$ decreases with the increase of eccentricity ratio (ε) with the change in

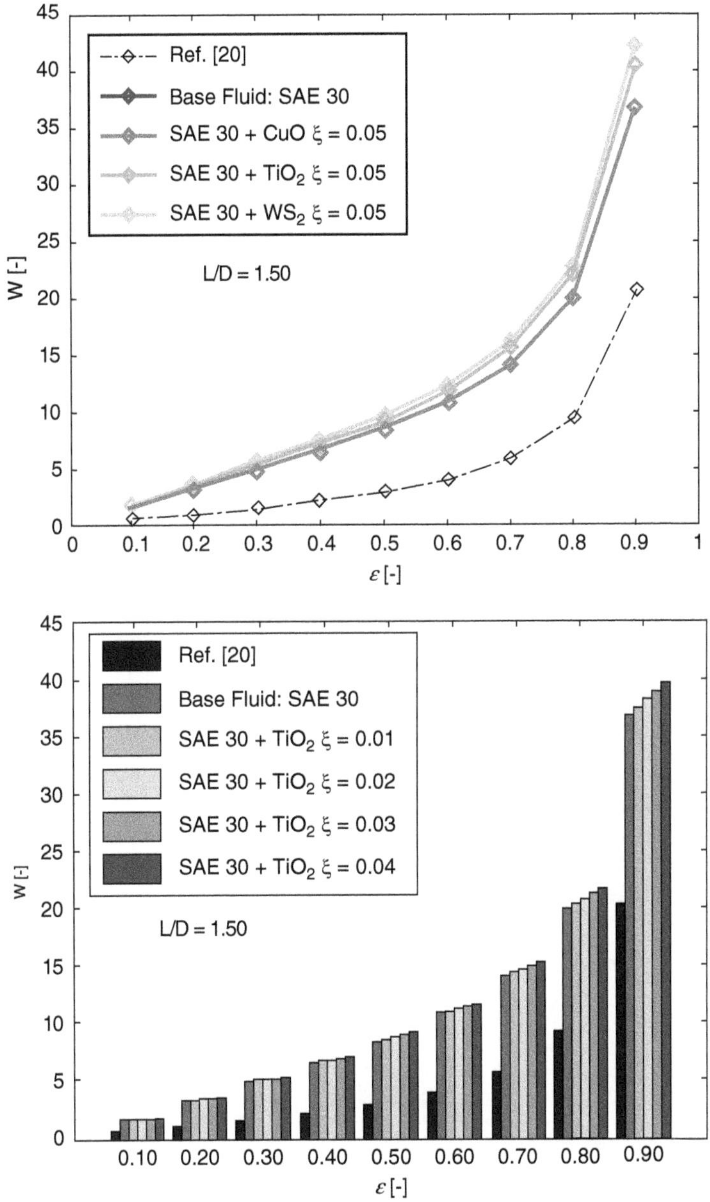

Figure 8.3 (a) Deviation of $\bar{W}$ with ε for base fluid and CuO, TiO_2 and WS_2 nano-lubricants. (b) Deviation of $\bar{W}$ with ε for base fluid and selected volume fractions of TiO_2.

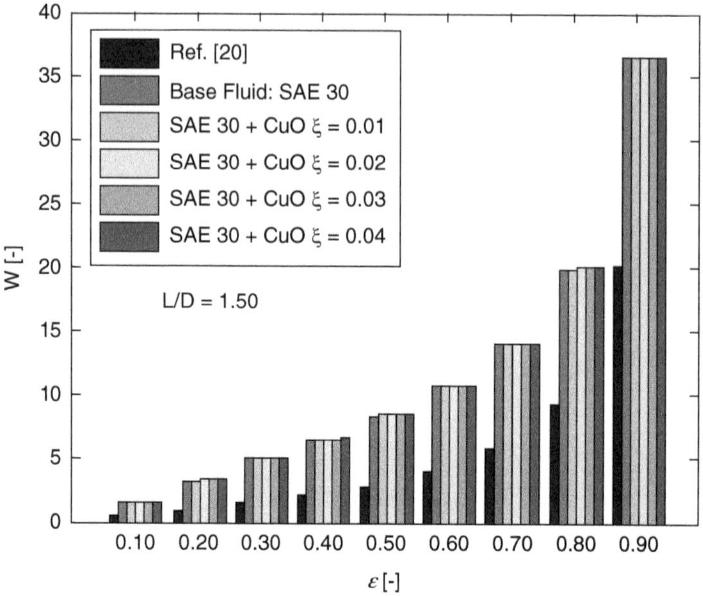

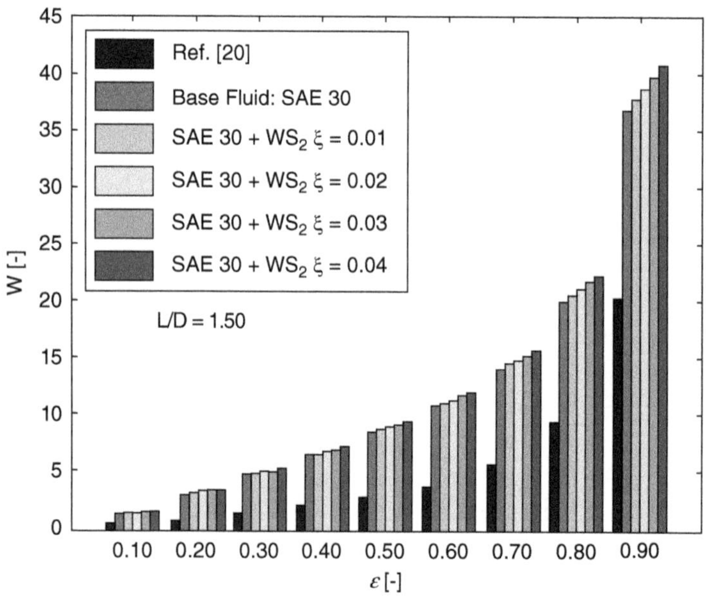

Figure 8.4 (a) Deviation of $\bar{W}$ with ε for base fluid and selected volume fractions of CuO. (b) Deviation of $\bar{W}$ with ε for base fluid and selected volume fractions of WS_2.

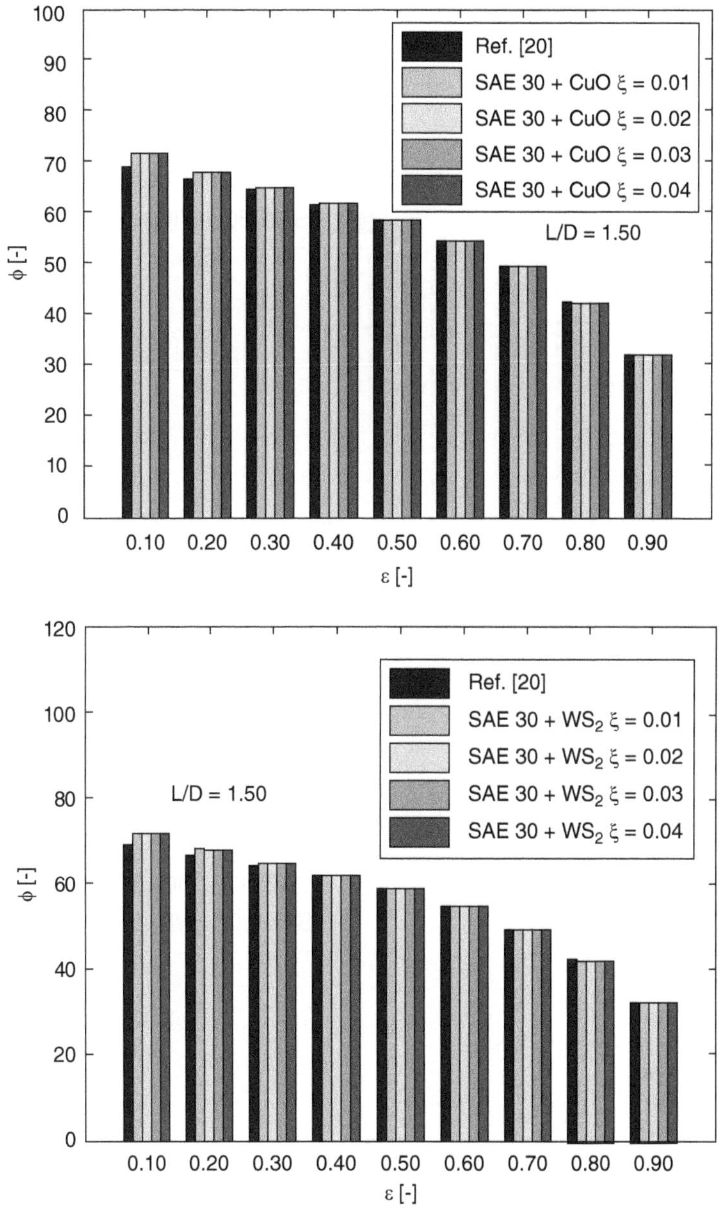

Figure 8.5 (a) Deviation of ϕ with ε for selected volume fractions of CuO. (b) Deviation of ϕ with ε for selected volume fractions of WS_2.

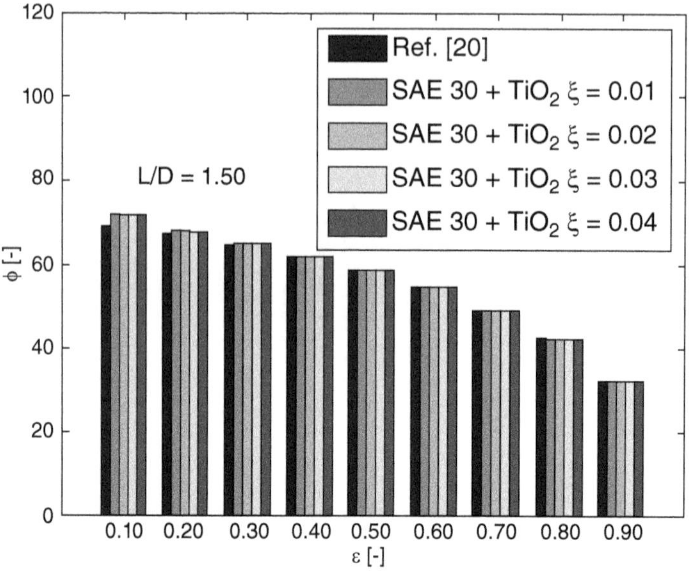

Figure 8.6 Deviation of ϕ with ε for selected volume fractions of TiO_2.

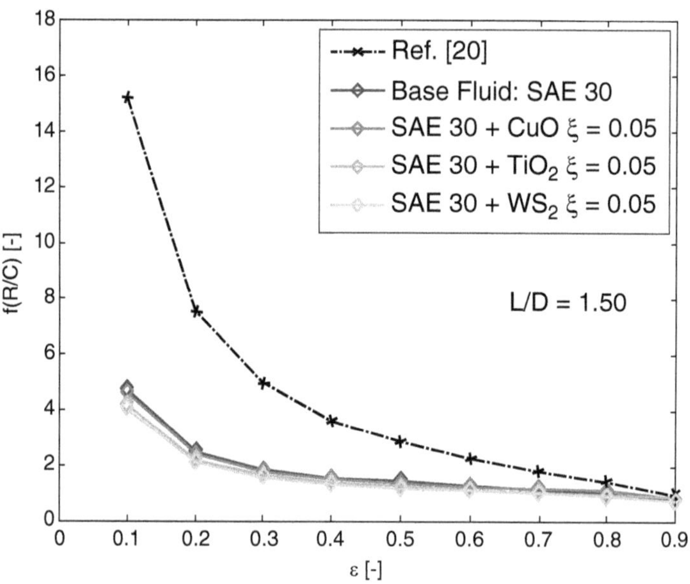

Figure 8.7 Deviation of $f(R/C)$ with ε for base fluid and CuO, TiO_2 and WS_2 nano-lubricants.

nanoparticles in the base fluid. It is also recognized that the fluid-film bearings lubricated with nano-lubricants (CuO, TiO_2, and WS_2 nanoparticles at $\xi = 0.05$) exhibit lesser $f(R/C)$ than the base lubricant (SAE 30).

8.6 CONCLUSIONS

This study compares the performance of a long bearing that operates with base lubrication and nano-lubricants at varying concentrations of nanoparticles. The following conclusions can be made in light of the findings.

1. In the nano-lubricants flow through the bearing, the circumferential pressure and maximum pressure in the lubricant film increase considerably with an increment in the volume fraction of nanoparticles in the nano-lubricant with the base fluid.
2. As the volume fraction of nanoparticles in the nano-lubricant with the base fluid increases, the load capacity of the journal bearing also increases significantly in the nano-lubricant flow regime.
3. The curtailment in the bearing attitude angle is monitored in the nano-lubricant flow regime.
4. The friction coefficient variable of the nano-lubricant flow regime wanes with definite values of the volume fraction of nanoparticles in the nano-lubricant with the base fluid.
5. The variation of the volume fraction of nanoparticles in the nano-lubricant with the base fluid affects the rheological performance of the long journal bearing quite significantly.

NOMENCLATURE

c	Radial clearance, m
e	Bearing eccentricity, m
R	Radius of journal, m
L	Length of bearing, m
D	Diameter of journal, m
p	Fluid-film pressure, N/m^2
$\mu_{BaseFluid}$	Absolute viscosity of base fluid, N-s/m^2
$\mu_{NanoFluid}$	Absolute viscosity of nano fluid, N-s/m^2
h	Oil film thickness, m
U	Velocity of journal, m/s

DIMENSIONLESS PARAMETERS

$\bar{p}$	Fluid-film pressure
$\bar{H}$	Oil film thickness
ξ	Volume fraction

ε	Eccentricity ratio = e / c
θ	Circumferential coordinate measured from the journal center
$\bar{\mu}$	Lubricant viscosity = $(\mu_{NanoFluid} / \mu_{BaseFluid})$
n	Number of nodes in each element
N_j^p	Shape functions for pressure at node j
$\bar{W}_r$	Oil film force along the line of centers
$\bar{W}_\theta$	Oil film force perpendicular to the line of centers
$\bar{W}$	Resultant load capacity
φ	Attitude angle (deg.)
$f\left(\frac{R}{C}\right)$	Friction coefficient variable
$\bar{F}_j$	Frictional force

MATRICES

$\left[\bar{F}_M\right]$	Fluidity matrix for the entire assembly
$\{\bar{p}\}$	Sub-matrix for pressure terms for element equation
$\{\bar{Q}_V\}$	Sub-matrix for flow terms for element equation
$\{\bar{R}_H\}$	Column matrix for nodal pressure variable for element equation
b	Column matrix for right-hand side variables for the entire assembly

REFERENCES

(1) Dowson, D. 1962. A generalized reynolds equation for fluid-film lubrication. *International Journal of Mechanical Sciences*, **4**: 159–70.
(2) Wada, S., Hirotsugu, H. and Masakazu, M. 1971. Application of finite element method to hydrodynamic lubrication problem. *Bulletin of Japan Society of Mechanical Engineers*, **14**: 1222–33.
(3) Li, B., Wang, X., Liu, W. and Xue, Q. 2006. Tribo-chemistry and antiwear mechanism of organic-inorganic nanoparticles as lubricant additives. *Tribolology Letters*, **22** (1): 79–84.
(4) Chen, H., Ding, Y. and Tan, C. 2007. Rheological behaviour of nano-fluids. *New Journal of Physics*, **9** (10): 367.
(5) Battez, A. H., Gonzalez, R., Felgueroso, D., Fernandez, J. E., Fernandez, M. R., Garcia, M. A. et al. 2007. Wear prevention behaviour of nanoparticle suspension under extreme pressure conditions. *Wear*, 263: 1568–74.
(6) Rico, E. F., Minondo, I. and Cuervo, D. G. 2007. The effectiveness of PTFE nanoparticle powder as an EP additive to mineral base oils. *Wear*, 262: 1399–1406.
(7) Peng, Y., Hu, Y. and Wang, H. 2007. Tribological behaviours of surfactant functionalized carbon nanotubes as lubricant additive in water. *Tribolology Letters*, **25**(3): 247–53.

(8) Wu, Y. Y., Tsui, W. C. and Liu, T. C. 2007. Experimental analysis of tribological properties of lubricating oils with nanoparticle additives. *Wear*, 262: 819–25.

(9) Joly-Pottuz, L., Vacher, B., Mogne, T. L., et al. 2008. The role of Nickel in Ni-containing nanotubes and onions as lubricant additives. *Tribolology Letters*, **29**(3): 213–19.

(10) Nair, K. P., Ahmed, M. S. and Al-Qahtani, S. T. 2009. Static and dynamic analysis of hydrodynamic journal bearing operating under nano-lubricants. *International Journal Of Nanoparticles*, **2** (1–6): 251–62.

(11) Mosleh, M., Atnafu, N. D., Belk, J. H. and Nobles, O. M. 2009. Modification of sheet metal forming fluids with dispersed nanoparticles for improved lubrication. *Wear*, **267**: 1220–25.

(12) Peng, D. X., Kang, Y., Hwang, R. M., Shyr, S. S. and Chang Y. P. 2009. Tribological properties of diamond and SiO2 nano-particles added in paraffin. *Tribolology International*, **42**: 911–17.

(13) Lee, K., Hwang, Y., Cheong, S. et al. 2009. Understanding the role of nanoparticles in nano-oil lubrication. *Tribology Letters*, **35**: 127–31.

(14) Madhusree, K. and Dey T. K. 2011. Effect of aggregation on the viscosity of copper oxide gear oil nano-fluids. *International Journal of Thermal Sciences*, **50**: 1741–47.

(15) Li, W., Zheng, S., Cao, B. and Ma, S. 2011. Friction and wear properties of ZrO2/SiO2 composite nanoparticles. *Journal of Nanoparticle Research*, **13**(5): 2129–37.

(16) Shenoy, B. S., Binu, K. G., Pai, R., Rao, D. S. and Pai, R. S. 2012. Effect of nanoparticles additives on the performance of an externally adjustable fluid film bearing. *Tribolology International*, **45** (1): 38–42.

(17) Shenoy, B. S., Binu, K. G., Pai, R., Rao, D. S. and Pai, R. S. 2014. Static characteristics of a fluid film bearing with TiO2 based nano-lubricant using the Modified Krieger–Dougherty viscosity model and couple stress model. *Tribology International*, **75**: 69–79.

(18) Binu, K. G., Shenoy, B. S., Rao, D. S. and Pai, R. 2014. A variable viscosity approach for the evaluation of load carrying capacity of oil lubricated journal bearing with TiO2 nanoparticles as lubricant additives. *Procedia Materials Science*, **6**: 1051–67.

(19) Kalakada, S. B., Nair, K. P. and Rajendrakumar, P. K. 2014. Computational analysis of journal bearing operating under lubricant containing Al2O3 and ZnO nanoparticles. *International Journal of Engineering, Science and Technology*, **6**(1): 34–42.

(20) Khonsari, M. M. and Booser, E. R. 2017. *Applied Tribology: Bearing Design and Lubrication*. Third edition. Hoboken, New York: John Wiley & Sons Inc.

(21) Wei, D., Bassem, K., Hong, G. and Hong, L. 2016. Roles of nanoparticles in oil lubrication. *Tribology International*, **102**: 88–98.

(22) Selvakumar, R. D. and Dhinakaran, S. 2017. Effective viscosity of nanofluids – A modified Krieger–Dougherty Model based on particle Size distribution (PSD) analysis. *Journal of Molecular Liquids*, **225**: 20–27.

(23) Harnoy, A. 2003. *Bearing Design in Machinery: Engineering Tribology and Lubrication*. Boca Raton: CRC Press.

Chapter 9

Tribological studies on Ti_6Al_4V surface

Priyam Kashyap, Suraj Tiwari, Santosh Kumar Tamang, and R. Jose Immanuel

9.1 INTRODUCTION TO TRIBOLOGY

Tribology, a captivating scientific realm originating from the Greek word "tribos," meaning "to rub," immerses itself in the intricate dance of interacting surfaces in relative motion. At its core, tribology addresses the fundamental concerns of friction, wear, and lubrication, revealing profound implications for various aspects of our world. From the seamless operation of car engines to the precision of medical prosthetics, tribology is the silent force that underpins the performance and durability of materials and devices. The science and engineering of interacting surfaces in relative motion form the bedrock of tribology. This multidisciplinary field encompasses the comprehensive study of friction, wear, lubrication, and related phenomena, spanning crucial applications in biomedical engineering, energy production, manufacturing, and transportation. A notable challenge within tribology lies in understanding and enhancing the tribological properties of titanium alloys, particularly significant for their widespread use in biomedical implants. These alloys boast biocompatibility, corrosion resistance, and a low elastic modulus, yet they are hampered by poor wear resistance and a high-friction coefficient, limiting their potential in wear-related applications. The unsung hero in the battle against friction and wear, lubrication, involves introducing a fluid or solid intermediary between interacting surfaces. By separating asperities and reducing energy dissipation, lubrication can dramatically enhance efficiency and extend component life. From humble grease in bearings to sophisticated nano-lubricants in MEMS devices, the spectrum of lubricants is as diverse as the applications they serve. Recent advancements in tribology have further blurred the lines between lubrication and surface engineering, with the development of self-lubricating surfaces and triboelectric effects offering exciting possibilities for minimizing friction and wear. Tribology reaches far beyond these core tenets, encompassing diverse subfields, such as stereochemistry, bio-tribology, and nanotribology. Its profound impact pervades a vast array of disciplines, from mechanical engineering and materials science to medicine and environmental science. By harnessing the principles of tribology, we can unlock countless possibilities for optimizing efficiency, prolonging lifespan, and minimizing

 DOI: 10.1201/9781003717362-9

energy consumption across a multitude of technological frontiers. This investigates the influence of various bio-lubricants, including deionized water, physiological saline, and bovine serum, on the friction and wear behavior of Ti_6Al_4V alloy, a prevalent titanium alloy. Through rigorous exploration, the study aims to unravel the wear mechanisms inherent in Ti_6Al_4V alloy under diverse lubrication conditions. A driving force behind tribological advancements is the escalating emphasis on sustainable practices. Traditional lubricants, often derived from fossil fuels, evoke environmental concerns due to their nonbiodegradable nature and potential toxicity. In response, researchers have redirected their focus toward eco-friendly alternatives, exemplified by bio-lubricants, to mitigate these issues. The significance of sustainable tribology resonates with its potential to usher in a greener future for mechanical systems. Bio-lubricants, derived from renewable sources, such as vegetable oils and animal fats, bring intrinsic benefits, such as biodegradability, reduced toxicity, and a smaller carbon footprint. This paradigm shift aligns seamlessly with the global commitment to curbing the environmental impact of industrial processes. As this exploration of tribology unfolds, the subsequent chapters will delve deeper into the principles of bio-lubrication, scrutinize its performance characteristics, and illuminate its applications. Simultaneously, a comprehensive examination of the tribological properties of Ti_6Al_4V will be conducted, contributing to a multidimensional analysis. The overarching goal of this book is to enrich the discourse on sustainable tribology, paving the way toward more efficient and environmentally friendly mechanical systems. The physical appearance of Ti_6Al_4V is seen in Figure 9.1.

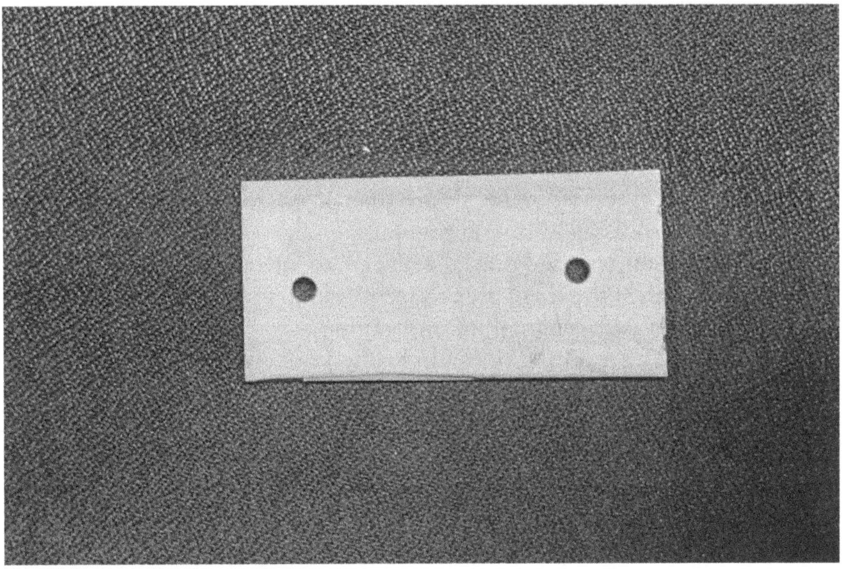

Figure 9.1 Physical appreances of Ti_6Al_4V plate with cut out with the help of wire cut EDM.

9.2 INTRODUCTION TO TI_6AL_4V

9.2.1 Titanium ore

Titanium ore refers to natural deposits or rocks containing titanium-bearing minerals from which titanium can be extracted. Titanium is a chemical element with the symbol Ti and atomic number 22. It is a lustrous transition metal with silver color, low density, and high strength. Titanium is commonly used in various industrial applications due to its excellent corrosion resistance, high strength-to-weight ratio, and biocompatibility. The primary minerals containing titanium are known as titanium ores (Subasinghe & Ratnayake, 2022).

The most common titanium ores include the following:

Ilmenite ($FeTiO_3$): Ilmenite is the most abundant titanium ore. It contains iron, titanium, and oxygen and is often found in beach sands, particularly in countries with heavy mineral sand deposits.

Rutile (TiO_2): Rutile is another significant source of titanium. It has a higher titanium content compared to ilmenite and is often found in metamorphic rocks and beach sands.

Leucoxene: Leucoxene is an altered form of ilmenite and rutile. It contains higher levels of iron and is often found in weathered mineral deposits.

Anatase (TiO_2): While less common, anatase is another titanium mineral. It is often found in association with rutile and is one of the three naturally occurring forms of titanium dioxide.

The extraction of titanium from these ores involves several steps:

Mineral Processing: The titanium ore is first mined and then subjected to mineral processing techniques to separate the valuable minerals from the gangue (unwanted material).

Smelting: The extracted minerals are then processed through smelting or chemical processes to obtain titanium dioxide (TiO_2).

Reduction: Titanium dioxide is reduced to metallic titanium using a method such as the Kroll process or the Hunter process, which involves the reaction with magnesium or sodium.

Purification: The resulting titanium is often further purified to meet specific industrial or aerospace requirements.

It is worth noting that the extraction and processing of titanium can be complex and energy-intensive. The availability and economic feasibility of titanium extraction depend on the concentration of titanium in the ore, the cost of extraction, and the demand for titanium in various industries.

As of my knowledge cutoff in January 2022, major titanium-producing countries include China, Australia, South Africa, Canada, and India (Jena et al., 1995; *Minerals | Free Full-Text | Titanium: An Overview of Resources and Production Methods*, n.d.).

Historical Evolution: The history of Ti_6Al_4V, a titanium alloy, dates back to the mid-20th century. The story of Ti_6Al_4V begins in the 1930s, with the discovery of titanium's potential in aircraft applications. However, its high reactivity and melting point posed significant challenges. Developed during the 1950s, this alloy gained prominence due to its exceptional combination of strength, corrosion resistance, and biocompatibility. Originally known by the trade name "Ti-6-4," it was primarily utilized in aerospace applications, particularly in the construction of aircraft components, where its lightweight yet sturdy characteristics proved invaluable. Over the years, Ti_6Al_4V has found expanded applications beyond aerospace, becoming a cornerstone material in various fields, including medical implants, automotive engineering, and sports equipment. Its journey from aerospace engineering to diverse sectors underscores the versatility and adaptability that define this titanium alloy.

Composition and Structure: Ti_6Al_4V is an alpha–beta titanium alloy, primarily composed of titanium (Ti), aluminum (Al), and vanadium (V). The specific composition is approximately 90% titanium, 6% aluminum, and 4% vanadium, with trace amounts of other elements, such as iron and oxygen, as shown in Table 9.1 and the graph shown in Figure 9.2. This alloy's unique combination of elements imparts superior mechanical properties, making it a preferred material in demanding applications. The alloy's dual-phase microstructure consists of both alpha and beta phases, contributing to its remarkable strength and corrosion resistance. This structural configuration allows Ti_6Al_4V to withstand harsh environments, including corrosive atmospheres and high-stress conditions, making it particularly suitable for critical applications in various industries (Figure 9.3).

Table 9.1 Composition chart investigated through EDX

Element	*Weight %*	*MDL*	*Atomic %*	*Net Int.*	*Error %*	*R*	*A*	*F*
Al K	5.2	0.13	9	324.8	7.2	0.8681	0.5357	1.0109
Ti K	90.6	0.73	87.2	2499.1	2.4	0.9149	0.9663	1.0134
V K	4.2	1.19	3.8	96.4	7.9	0.9201	0.9594	1.0163

Figure 9.2 Titanium Ore and Pen for size reference.

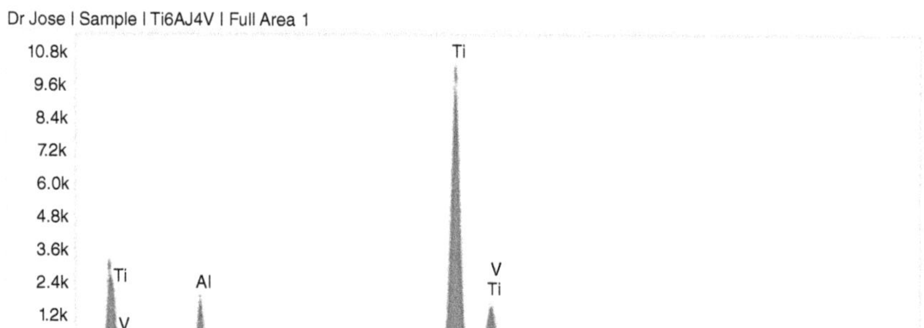

Figure 9.3 EDX graph of Ti_6Al_4V sample.

9.2.2 Microstructure of titanium

The microstructure of Ti_6Al_4V, a widely used titanium alloy, plays a crucial role in determining its mechanical properties and performance. There are four types of microstructure of Ti6 alloy, namely, a, b, c, and d, respectively, as shown in Figure 9.4 (Attanasio et al., 2013). Ti_6Al_4V is an alpha–beta alloy, meaning it contains both alpha (α) and beta (β) phases at room temperature. Here is an overview of the microstructure of Ti_6Al_4V:

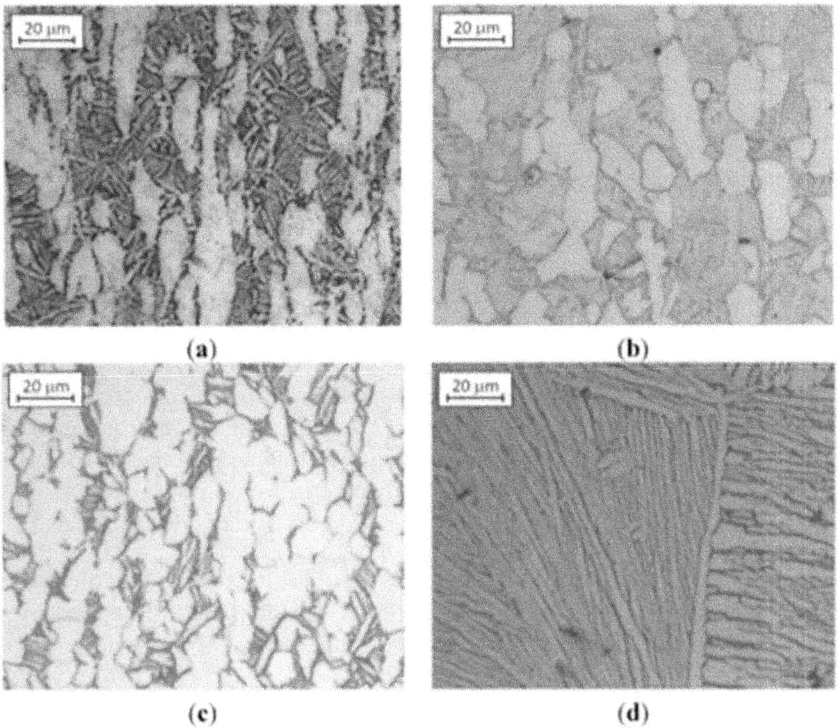

Figure 9.4 (a) Mill annealed; (b) bimodal; (c) fully equiaxed and (d) fully lamellar.

Alpha Phase (α): The alpha phase in Ti_6Al_4V is characterized by a hexagonal close-packed (hcp) crystal structure. This phase provides good strength, toughness, and corrosion resistance.

Beta Phase (β): The beta phase has a body-centered cubic (bcc) crystal structure. The beta phase in Ti_6Al_4V contributes to the alloy's high-temperature strength and processability.

Alpha–Beta Microstructure: Ti_6Al_4V is designed to have a dual-phase microstructure, containing both alpha and beta phases. The balance between these phases can be adjusted through heat treatment, allowing for the optimization of mechanical properties.

Heat Treatment Effects: The microstructure of Ti_6Al_4V is significantly influenced by heat treatment processes. Annealing, solution treatment, and aging are common heat treatments employed to achieve specific mechanical properties. Heat treatment controls the transformation between alpha and beta phases, grain size, and precipitation of secondary phases.

Grain Size: The grain size of Ti_6Al_4V is an important factor in determining mechanical properties. Fine-grained structures often result in improved strength and toughness. Grain size is influenced by the heat treatment process.

Secondary Phases and Inclusions: Depending on the alloy composition and processing conditions, Ti_6Al_4V may contain secondary phases and inclusions. The presence of elements like aluminum and vanadium can lead to the formation of precipitates, which can affect mechanical properties.

Texture: The microstructure of Ti_6Al_4V can exhibit certain preferred crystallographic orientations or textures. The texture can influence anisotropic properties, particularly in wrought or rolled materials.

Alpha Prime (α') Phase: In certain conditions, Ti_6Al_4V may exhibit a metastable martensitic phase known as alpha prime (α'). This phase is characterized by a body-centered orthorhombic crystal structure and can influence the mechanical properties of the alloy.

Understanding and controlling the microstructure of $Ti\ {}_6Al_4V$ are essential for tailoring the alloy's properties to meet specific application requirements. Microstructural analysis techniques, including metallography, electron microscopy, and X-ray diffraction, are commonly used to study and characterize the microstructure of Ti_6Al_4V and ensure the desired material performance.

Mechanical Properties: Ti_6Al_4V is renowned for its outstanding mechanical properties, making it one of the most widely used titanium alloys (Table 9.2). Some key characteristics include:

- High Strength-to-Weight Ratio: Ti_6Al_4V boasts an impressive strength-to-weight ratio, making it exceptionally strong while remaining lightweight. This property is especially critical in aerospace applications where weight reduction is a primary consideration.
- Corrosion Resistance: The alloy's resistance to corrosion, particularly in aggressive environments, enhances its durability and lifespan. This property makes Ti_6Al_4V a preferred choice for biomedical implants and components exposed to corrosive substances.
- Biocompatibility: Recognized for its excellent biocompatibility, Ti_6Al_4V is extensively used in medical applications, including orthopedic implants and dental prosthetics. Its compatibility with the human body minimizes the risk of adverse reactions.
- Heat Resistance: Ti_6Al_4V exhibits good heat resistance, maintaining its structural integrity at elevated temperatures. This characteristic is crucial for applications in aerospace and industrial settings exposed to high-temperature conditions.

Applications: The versatility of Ti_6Al_4V has led to its widespread adoption across diverse industries:

- Aerospace: Used in aircraft components, landing gear, and structural elements due to its high strength and lightweight nature.
- Medical: A preferred material for orthopedic implants, dental implants, and surgical instruments owing to its biocompatibility and corrosion resistance.

Table 9.2 Properties of Ti_6Al_4V

Property	*Value*	*Units*
Density	4.43	g/cc
Melting point	1604–1660	°C
Solidus	1604	°C
Liquidus	1660	°C
Beta Transus	980	°C
Tensile strength, ultimate	1170	MPa
Tensile strength, yield	1100	MPa
Elongation at break	10	%
Modulus of elasticity	114	GPa
Brinell hardness	334	HB
Rockwell hardness (C)	363	HRC
Vickers hardness	36	HV
Electrical resistivity	178	μΩcm
Poisson's ratio	0.342	

- Automotive: Employed in automotive components, such as exhaust systems and engine parts, benefiting from its strength and resistance to corrosion.
- Sports Equipment: Found in bicycle frames, golf clubs, and other sports equipment, where the combination of strength and lightness is crucial.

9.3 TRIBOLOGY AND SUSTAINABILITY: FORGING A GREENER PATH IN MECHANICAL SYSTEMS

Tribology, the intricate study of friction, wear, and lubrication, transcends its mechanical origins to become a vital nexus between engineering and sustainability. This section delves into the evolving landscape of tribology, exploring its pivotal role in the quest for sustainable practices within various industries. Tribology stands at the forefront of a paradigm shift toward sustainable practices. Traditional lubricants, often derived from fossil fuels, have long raised environmental concerns due to their nonbiodegradable nature and potential toxicity. In response, the field of tribology is steering away from conventional solutions, propelling an industry-wide transition toward environmentally conscious alternatives. The environmental ramifications of traditional lubricants necessitate a reevaluation of their usage. From industrial machinery to automotive engines, the pervasive use of these lubricants contributes to pollution and poses challenges for ecosystems. Tribologists are actively investigating the ecological footprint of these lubricants, laying the groundwork for a more sustainable future.

- **Bio-lubricants: An Overview:**
  Bio-lubricants emerge as beacons of sustainability within the realm of tribology. Derived from renewable resources, such as vegetable oils and animal fats, these alternatives present a compelling case for reducing dependency on finite fossil fuels. Their intrinsic benefits, including biodegradability, reduced toxicity, and a smaller carbon footprint, make them pivotal players in sustainable tribological practices. As industries worldwide grapple with the need to adopt eco-friendly practices, the significance of sustainable tribology becomes increasingly pronounced. By actively exploring alternative lubrication methods and materials, tribologists are not only mitigating environmental impact but also contributing to the broader commitment to sustainability. This section sets the stage for a deeper exploration of bio-lubrication principles, their performance characteristics, and their applications in subsequent chapters, offering a comprehensive view of tribology's role in forging a greener and more sustainable future for mechanical systems.
- **Bio-lubrication Principles:** Paving the Way for Sustainable Mechanical Systems
  The exploration of bio-lubrication principles within the realm of tribology represents a crucial stride toward sustainable practices. This section delves into the core characteristics and principles that make bio-lubricants stand out as eco-friendly alternatives, elucidating their potential to revolutionize the field of mechanical systems.
- **Biodegradability and Less Toxicity:**
  Bio-lubricants shine as environmentally responsible choices due to their inherent biodegradability. Unlike traditional lubricants that may persist in the environment, bio-lubricants undergo natural decomposition processes, reducing the risk of long-term ecological impact. In addition, their reduced toxicity further diminishes potential harm to ecosystems, aligning with the broader goal of minimizing environmental footprints (Gnanasekaran & Chavidi, 2018).
- **Carbon Footprint Reduction:**
  A standout feature of bio-lubricants is their capacity to contribute to a smaller carbon footprint. Derived from renewable resources, these lubricants often undergo manufacturing processes with lower energy consumption and greenhouse gas emissions. Assessing and quantifying this reduction in carbon footprint solidifies the role of bio-lubrication in sustainable tribology (Singh et al., 2020).
- **Applications of Bio-lubrication:**
  Understanding the real-world applications of bio-lubricants is pivotal for envisioning their widespread adoption. This section explores the current uses of bio-lubrication in diverse mechanical systems. Case studies and success stories highlight instances where bio-lubricants have not only proven effective in maintaining machinery integrity but have also demonstrated their potential to outperform traditional lubricants in specific applications (Xu et al., 2019).

9.3.1 Performance characteristics of bio-lubricants: Bridging efficiency and environmental impact

In the quest for sustainable tribology, understanding the performance characteristics of bio-lubricants is paramount. This section navigates through the intricacies of bio-lubrication, conducting a comparative analysis with traditional lubricants. By exploring efficiency, effectiveness, and environmental impact, we unravel the potential of bio-lubricants to revolutionize mechanical systems while minimizing their ecological footprint.

- **Comparative Analysis with Traditional Lubricants:**
  This subsection scrutinizes the performance of bio-lubricants in comparison to their traditional counterparts. Through systematic studies, researchers have evaluated factors such as friction reduction, wear resistance, and overall lubrication efficacy. Comparative analyses provide insights into the strengths and limitations of bio-lubricants, serving as a foundation for informed decision-making in selecting lubrication solutions (Naveed et al., 2021).
- **Efficiency and Effectiveness:**
  Efficiency in tribological applications encompasses not only the lubricant's ability to reduce friction and wear but also its impact on overall system performance. This subsection delves into studies evaluating the efficiency and effectiveness of bio-lubricants in different mechanical systems. Understanding how these lubricants contribute to the longevity and reliability of machinery is crucial for their broader adoption (Katiyar et al., 2023).

9.4 ENHANCING TRIBOLOGICAL PROPERTIES OF TI_6AL_4V

To increase the tribological (friction, wear, and lubrication) performance of Ti_6Al_4V (Titanium alloy containing 6% aluminum and 4% vanadium), several strategies can be employed. These approaches aim to enhance the material's resistance to wear and reduce friction under different operating conditions. In order to characterize the tribological performance, a few tests are essential. Some of the noteworthy methods are as follows.

Morphological analysis: Morphological analysis involves a detailed examination of the structure and form of Ti_6Al_4V specimens, providing insights into their surface characteristics at a microscopic and, potentially, nanoscopic level. This analysis is crucial for understanding the topographical features, surface roughness, and any modifications resulting from the experimental processes. Techniques such as scanning electron microscopy (SEM) and atomic force microscopy (AFM) may be employed to capture high-resolution images, unveiling the intricate details of the Ti_6Al_4V morphology. In morphological analysis, initially, we took an FE-SEM image of a polished Ti6 sample, followed by the textured surface of Ti_6Al_4V.

Wettability Study: Wettability is a critical parameter that defines the interaction between a liquid and a solid surface. In the context of Ti_6Al_4V, understanding wettability is essential for evaluating its compatibility with various substances and for applications where surface interactions play a pivotal role. Techniques like the Kruss Drop Shape Analyzer are employed to measure contact angles, providing quantitative data on how liquids spread or bead on the Ti_6Al_4V surface.

Several methods were explored by researchers across the world to alter the surface properties of Ti_6Al_4V with a special focus on improved tribological attributes. Some of the critical methods are described below.

9.4.1 Surface texturing

Surface texturing of titanium refers to the process of modifying the surface of titanium materials to create specific patterns, structures, or textures. This can be done through various methods, such as chemical etching, laser processing, or mechanical techniques. The aim of surface texturing is often to enhance certain properties of the material, such as hydrophobicity. Surface texturing on the Ti_6Al_4V can be achieved using a high-speed micro-milling machine equipped with a carbide tool. This process involves the creation of micro-pillars through the tool's crisscross maneuvering, forming micro-channels on the Ti_6Al_4V surface. The resulting machining pattern leads to the development of micro-pillar structures on the Ti_6Al_4V surface, contributing to an increase in hydrophobicity and a subsequent decrease in wettability. The wettability of a surface plays a crucial role in various properties such as adhesion, friction, and corrosion, with a non-wetting surface being less susceptible to corrosion by water or other liquids. From Sharma & Grewal, 2023, it is evident that the studies demonstrate enhancements in frictional properties through bioinspired surfaces. Overall, the research explores innovative approaches to mitigate energy losses and improve tribological performance using bioinspired materials.

The surface texturing pattern drew inspiration from the microstructure of lotus leaves, showcasing improvements in frictional properties through bioinspired surfaces. This supports innovative approaches to reduce energy losses and enhance tribological performance using bioinspired materials.

Hydrophobicity refers to the ability of a surface to repel water. Achieving hydrophobicity on titanium surfaces is of interest in various applications, including medical devices, aerospace components, and outdoor equipment. Here are some common methods used to achieve hydrophobicity on titanium surfaces.

9.4.2 Chemical treatments

Fluorination: Treating titanium surfaces with fluorine-containing compounds can create a thin layer of hydrophobic material on the surface, making it water-repellent.

Self-Assembled Monolayers (SAMs): Organic molecules with hydrophobic properties can be chemically bonded to the titanium surface, creating a hydrophobic layer.

9.4.3 Surface roughening

Micro- or Nano-structuring: Introducing micro or nanostructures on the titanium surface can alter its wetting properties, making it more hydrophobic. This can be achieved through techniques, like laser ablation or acid etching. In some cases, people go for plasma treatment of the titanium surface. Subjecting the titanium surface to plasma treatment can modify its surface energy, making it more hydrophobic.

9.4.4 Coating techniques

Hydrophobic Coatings: Applying specific hydrophobic coatings, such as those containing fluoropolymers or siloxanes, can make the titanium surface water-resistant.

Nanostructured Coatings: Creating nanostructures on the titanium surface can enhance hydrophobicity. This can be achieved through methods like sol–gel processing or chemical vapor deposition.

9.4.4.1 Coating materials

- Diamond-Like Carbon (DLC): DLC coatings provide excellent hardness, low friction, and wear resistance. They are commonly used to reduce friction and wear on Ti_6Al_4V surfaces.
- Nitrides and Carbides: Titanium nitride (TiN), titanium carbide (TiC), and other nitride or carbide coatings can enhance hardness and wear resistance, improving the tribological properties of the Ti_6Al_4V nano-powders.
- Polymers and Lubricious Coatings: Some applications may benefit from the incorporation of polymer-based or lubricious coatings to reduce friction and enhance wear resistance.

9.4.4.2 Coating methods

- Physical Vapor Deposition (PVD): PVD methods, such as sputtering or evaporation, involve depositing thin films of coating materials onto the nano-powders' surface. Common coatings include nitrides, carbides, and DLC, which can enhance hardness and wear resistance.
- Chemical Vapor Deposition (CVD): CVD is another technique for depositing thin films on nano-powder surfaces. It involves chemical reactions to create a coating, and materials, such as diamond, DLC, or various carbides, can be applied for improved tribological properties.
- Electroplating: Electroplating can be used to deposit metallic or alloy coatings onto the Ti_6Al_4V nano-powders. This method can provide corrosion resistance and enhanced wear properties.

- Sol–Gel Coating: Sol–gel processes involve the application of a liquid precursor solution that undergoes a chemical reaction to form a solid coating. This method is versatile and can incorporate various additives to improve tribological performance.

9.4.5 Surface stressing

In a few applications, a compressive residual stress is imparted onto the material's surface, which can provide improved tribological performance. Shot peening is one among those processes that involves shooting small particles at the surface to induce compressive stresses, improving fatigue resistance and wear performance.

In modern applications, shots are replaced by lasers. Laser treatment can be used to precisely modify the microstructure of the surface, enhancing hardness and wear resistance.

9.4.6 Heat-controlled manufacturing

Controlled heat treatment can be applied to improve the microstructure, leading to increased hardness and wear resistance. However, this needs to be carefully optimized to avoid compromising other mechanical properties (Alvi et al., 2021). The exploration of electron beam powder bed fusion (EBPBF) as a favorable manufacturing method for Ti_6Al_4V alloy, particularly in aerospace applications, has gained significant attention. The study focuses on evaluating the dry sliding wear behavior of Ti_6Al_4V produced by EBPBF against steel and alumina counterballs across a temperature range from room temperature to 500 °C. The aim is to understand how EBPBF processing and microstructure influence the wear properties of Ti_6Al_4V, with the goal of expanding its application criteria in aerospace.

Key findings from the wear tests indicate a decrease in wear rate with increasing temperature. This phenomenon is attributed to the formation of a stable oxide glaze layer. The study sheds light on the sliding wear behavior at elevated temperatures, explores wear mechanisms, and investigates microstructural changes beneath the wear track of EBPBF Ti_6Al_4V alloy in the presence of steel and alumina counterbodies. The outcomes contribute valuable insights into the performance of Ti_6Al_4V alloy in sliding wear conditions across a range of temperatures.

9.4.7 Improved lubrication

Usage of solid lubricants, such as graphite or MoS2, into the system can reduce friction and wear. Luo et al. (2013) discuss the wear behavior of Ti_6Al_4V, a titanium alloy widely used in biomedical applications, with a focus on the influence of different lubrication conditions. The study aims to clarify the wear mechanisms of Ti_6Al_4V when subjected to dry friction, deionized water, physiological saline, and bovine serum as lubricants. Friction and wear tests were conducted between Ti_6Al_4V plates and Si3N4 balls on a multifunctional tester, and the results were analyzed with a particular emphasis on wear resistance and mechanisms.

Employ Advanced Lubrication Techniques: Utilize advanced lubrication methods, such as hydrodynamic or hydrostatic lubrication, to minimize direct metal-to-metal contact.

9.5 SPECIAL APPLICATION OF TI_6AL_4V AS POWDERS

Incorporating Ti_6Al_4V powder as reinforcement introduces a novel dimension to the application domain of the material. This approach may bolster mechanical strength, wear resistance, or other desirable attributes. The process involves dispersing Ti_6Al_4V powder within a matrix, potentially a polymer or another material, creating a composite with tailored properties.

9.6 CONCLUSION

The extensive investigation into strategies for enhancing the tribological properties of Ti6Al4V emphasizes the multifaceted nature of optimizing its performance across diverse operating conditions. The integration of various techniques, as highlighted in this discussion, forms a comprehensive framework for advancing the material's tribological capabilities. Utilizing methodologies such as SEM and AFM provides valuable insights into Ti_6Al_4V's surface characteristics, while wettability assessments conducted with tools like the Kruss Drop Shape Analyzer contribute to evaluating its compatibility with different substances. Surface texturing methods, including chemical etching, laser processing, and high-speed micro-milling, enhance hydrophobicity and reduce wettability, complemented by the application of hydrophobic coatings through chemical treatments like fluorination and the use of self-assembled monolayers (SAMs). The emergence of electron beam powder bed fusion (EBPBF) as a favorable manufacturing method, particularly in aerospace applications, demonstrates improved wear resistance attributed to the formation of a stable oxide glaze layer. Improved lubrication techniques, incorporating solid lubricants such as graphite or MoS_2, alongside advanced methods, like hydrodynamic or hydrostatic lubrication, play a crucial role in minimizing friction and wear. The utilization of Ti_6Al_4V powder as reinforcement within composite matrices introduces innovative possibilities for enhancing mechanical strength and wear resistance. In summary, these diverse strategies contribute to a holistic understanding of optimizing Ti_6Al_4V's tribological performance, holding significance for researchers and practitioners in materials science, manufacturing, and various industries. The potential advancements in aerospace applications, medical devices, and other fields underscore the broader goal of improving material efficiency, reducing environmental impact, and fostering innovation in materials engineering, aligning with a commitment to advancing technology and sustainability.

REFERENCES

Alvi, S., Neikter, M., Antti, M.-L., & Akhtar, F. (2021). Tribological performance of Ti_6Al_4V at elevated temperatures fabricated by electron beam powder bed fusion. *Tribology International*, *153*, 106658. https://doi.org/10.1016/j.triboint.2020.106658

Attanasio, A., Gelfi, M., Pola, A., Ceretti, E., & Giardini, C. (2013). Influence of material microstructures in micromilling of Ti_6Al_4V alloy. *Materials*, *6*(9), Article 9. https://doi.org/10.3390/ma6094268

Gnanasekaran, D., & Chavidi, V. P. (2018). Biodegradable, renewable, and eco-friendly vegetable oil: Lubricants. In D. Gnanasekaran & V. P. Chavidi (Eds.), *Vegetable Oil Based Bio-lubricants and Transformer Fluids: Applications in Power Plants* (pp. 29–47). Springer. https://doi.org/10.1007/978-981-10-4870-8_2

Jena, B. C., Dresler, W., & Reilly, I. G. (1995). Extraction of titanium, vanadium and iron from titanomagnetite deposits at pipestone lake, Manitoba, Canada. *Minerals Engineering*, *8*(1), 159–168. https://doi.org/10.1016/0892-6875(94)00110-X

Katiyar, J. K. K., Abdul Rani, A. M., Sulaiman, M. H., Barman, U., Masset, P. J., Rao, T. V. V. L. N. (2023). Recent advances of tribology in sustainable manufacturing. In Jitendra Kumar Katiyar, T. V. V. L. N. Rao, Ahmad Majdi Abdul Rani, Mohd Hafis Sulaiman, J. Paulo Davim (Eds.), *Tribology in Sustainable Manufacturing*. CRC Press.

Luo, Y., Yang, L., & Tian, M. (2013). Influence of bio-lubricants on the tribological properties of Ti_6Al_4V alloy. *Journal of Bionic Engineering*, *10*(1), 84–89. https://doi.org/10.1016/S1672-6529(13)60202-4

Minerals | Free Full-Text | Titanium: An Overview of Resources and Production Methods. (n.d.). Retrieved January 15, 2024, from https://www.mdpi.com/2075-163X/11/12/1425

Naveed, T., Zahid, R., Mufti, R. A., Waqas, M., & Hanif, M. T. (2021). A review on tribological performance of ionic liquids as additives to bio lubricants. *Proceedings of the Institution of Mechanical Engineers, Part J: Journal of Engineering Tribology*, *235*(9), 1782–1806. https://doi.org/10.1177/1350650120973805

Sharma, S. K., & Grewal, H. S. (2023). Tribological behavior of bioinspired surfaces. *Biomimetics*, *8*(1), Article 1. https://doi.org/10.3390/biomimetics8010062

Singh, Y., Sharma, A., Singh, N., Singla, A., & Rastogi, P. M. (2020). Prospects of inedible plant oil-driven bio-lubricants for tribological characteristics – a review. *International Journal of Ambient Energy*, *41*(13), 1534–1547. https://doi.org/10.1080/01430750.2018.1517684

Subasinghe, H. C. S., & Ratnayake, A. S. (2022). General review of titanium ores in exploitation: present status and forecast. *Comunicações Geológicas*, *109*(1), 21–31. https://doi.org/10.34637/aab4-mk81

Xu, Y., Liu, Z., Dearn, K. D., Dong, Y., You, T., & Hu, X. (2019). Thermo-tribological behaviour of microgels for improved aqueous lubrication for steel/UHMWPE contact. *Tribology International*, *130*, 63–73. https://doi.org/10.1016/j.triboint.2018.08.039

Chapter 10

Bio-surface coatings for healthcare applications

R. Prayer Riju, S. Arulvel, D. Dsilva Winfred Rufuss, Jayakrishna Kandasamy, and P. Jeyapandiarajan

10.1 INTRODUCTION

In the domain of healthcare, bio-surface coatings play a pivotal role in terms of advantages such as biocompatibility, controlled release of active substances, and antimicrobial features. Coatings are crafted to alter the surface characteristics, catering to specific needs within biomedical contexts of various substrates, ranging from metal implants to nanoparticles. Recent research has delved into the synthesis, characterization, and efficacy of these coatings, both in controlled environments and within living organisms. Moreover, there is a growing interest in advancing techniques for applying bioactive coatings to medical devices. A notable focus has been on the development of antimicrobial coatings to ease the potential contribution of inanimate surfaces to the spread of infections in healthcare settings. An important aspect receiving attention from researchers is the selection of materials for bio-surface coatings. The material selection for the coatings is guided by several factors, such as biocompatibility, mechanical strength, and resistance to corrosion within the human body.

As we navigate this chapter, the significance of coating techniques and material selection in bio-surface coatings becomes apparent. This chapter embarks on a comprehensive exploration of bio-surface coatings for healthcare applications, delving into the multifaceted landscape of surface modification for polymer, metallic, and ceramic coatings. To set the stage, we begin with an insightful background, contextualizing the imperative need for bio-surface coatings in contemporary healthcare scenarios. As we navigate through the chapter, a key focus emerges on the classification of bio-surface coatings, from antimicrobial coatings to those enhancing biocompatibility. In particular, the classification unfolds a world of possibilities for tailored medical solutions. The subsequent sections illuminate the materials utilized in bio-surface coatings, offering a meticulous examination of their properties and applicability in healthcare settings. From biopolymer coatings to advanced nanomaterials, the selection of materials is scrutinized for their ability to meet the stringent demands of biomedical applications. The exploration does not conclude without a deep dive into coating techniques,

DOI: 10.1201/9781003717362-10

unravelling the methodologies employed in applying these bio-surface coatings. This chapter will serve as a foundational resource, guiding researchers, practitioners, and enthusiasts through the dynamic landscape of bio-surface coatings in healthcare applications

10.1.1 Background

The historical trajectory of bio-surface coatings has witnessed significant evolution, with notable milestones spanning from the 19th century to the present (Figure 10.1). In the early 18th century, the developments were primarily focused on the basic protective coatings for the metal surfaces. After this era, the mid-20th century showed a paradigm shift with the rise of polymer-based coatings, particularly the advent of epoxy coatings in the 1950s, marking a pivotal moment in protective coating technology. The latter half of the 20th century witnessed an expansion of research into biomedical applications, emphasizing biocompatibility (Bauer and Lieleg 2023). As the 21st century unfolded, the field experienced a surge in advancements pertaining to bio-surface coatings for medical devices and implants. Antimicrobial coatings gained prominence for their potential to mitigate infections. Furthermore, the progress in 2010 has shown the development of multifunctional coatings that incorporated biocompatibility, corrosion resistance, and antimicrobial properties (Pietsch et al. 2020). The integration of nanotechnology also played a pivotal role, facilitating the creation of advanced coatings with precise functionalities. While this historical overview provides insights into the evolution of bio-surface coatings till 2023, continued research beyond this point is necessary for a comprehensive understanding of recent developments and emerging trends in the field (Choi et al. 2020).

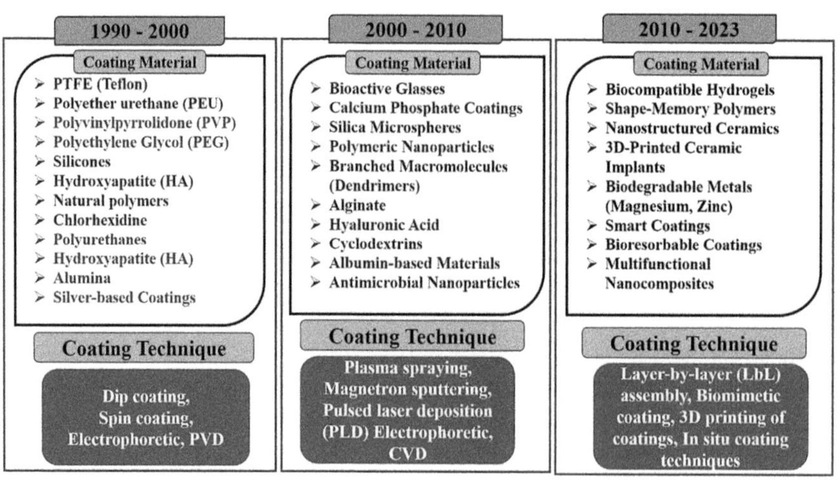

Figure 10.1 Historical evolution road map of bio-surface coating.

This evolution can be delineated across distinct stages. In the early developmental phase, the primary focus was on enhancing biocompatibility and bio-integration to improve the overall performance of medical devices and implants. Subsequently, as knowledge deepened regarding healthcare-associated infections (HAIs) and their transmission in healthcare settings, bio-surface coatings began incorporating antimicrobial properties to mitigate the risk of infections. After several years, the antimicrobial coating was not limited to the implant, but it is also used on medical devices. The escalating use of medical devices underscored the importance of corrosion resistance, emerging as a crucial property to ensure the longevity and efficacy of these devices. Progressing further, the multifunctional coatings were devised, amalgamating various properties such as biocompatibility, antimicrobial activity, and corrosion resistance (Choi et al. 2020). The advent of advanced deposition techniques, including atomic layer deposition and chemical vapour deposition, played a pivotal role in facilitating the uniform and durable application of bio-surface coatings across diverse substrates. As understanding deepened, clinical trials and in vivo studies were initiated to substantiate the efficacy and safety of bio-surface coatings in healthcare settings.

Research efforts have been directed towards the creation of bimolecular surface coatings, emphasizing properties such as bio-integration, corrosion resistance, and antimicrobial activity. Nanotechnology has recently played a crucial role in the advancement of antimicrobial coatings, providing potent applications for the control and eradication of pathogenic microorganisms in healthcare environments (Bauer and Lieleg 2023). Furthermore, advancements in biopolymer coatings have demonstrated outstanding potential across various biomedical applications, offering flexible functionalization, and the ability to modify surface properties to meet specific requirements in healthcare settings (Kışla et al. 2023).

10.2 CLASSIFICATION OF BIO-SURFACE COATINGS BASED ON THEIR PROPERTIES

Bio-surface coatings are thin layers applied to a surface that do not interact, respond to, or promote a chemical reaction or biological response. There are different types of bio-surface coatings, including bio-inert coatings, which do not interact with biological systems, and biomaterial coatings, which are used in biomedical and biological applications. The classification of bio-surface coatings is an important aspect that enables a systematic understanding of their diverse properties and functionalities. Bio-coatings, designed for healthcare applications, can be categorized based on antibacterial, corrosion resistance, drug-elucidating, adhesion, hydrophilic, and hydrophobic properties.

10.2.1 Antimicrobial coating

Antimicrobial coatings comprise the application of chemical agents on the surfaces to prevent the proliferation of disease-causing microorganisms, such as bacteria, fungi, and mould. These coatings find application in diverse healthcare applications, such as implants, drug delivery systems, and medical instruments. Several examples of antimicrobial coatings are notable and discussed as follows. The chlorhexidine-incorporated hydroxyapatite coating delivers both antimicrobial and biocompatibility properties, rendering them fitting for medical devices and implants. Furthermore, the combination of chlorhexidine with polymer and calcium phosphate has improved antimicrobial resistance. Quaternary ammonium-based coatings utilize compounds such as benzalkonium chloride or didecyldimethylammonium chloride to inhibit microorganism growth. Compounds like quaternary ammonium compounds, such as benzalkonium chloride or didecyldimethylammonium chloride, serve as active ingredients in antimicrobial coatings, effectively inhibiting the growth of microorganisms (Mubeen et al. 2021). At present, the silver-based antimicrobial coatings are widely employed due to their exceptional effectiveness in comparison to other types. This is because the silver acts as an active ingredient compared to the agents like titanium dioxide, zinc-based compounds, and other coatings discussed above.

10.2.2 Drug-eluting coatings

Drug-eluting coatings serve a crucial function by delivering drugs or active molecules to specific locations in the body, offering targeted therapy, and mitigating the risk of adverse side effects. Common materials used in these coatings encompass polymers, known for their versatility and the ability to be surface-modified, gradually dissolving over time when their function is fulfilled, allowing for controlled drug release. Responsive coatings like polymer brushes, which alter their properties in response to environmental changes, such as temperature, pH, and ionic strength, are employed to produce drug-eluting coatings that adapt to specific conditions. Polyelectrolyte multilayers, formed through layer-by-layer deposition of polycations and polyanions, provide another avenue for producing coatings that release drugs or active peptides. An illustrative application of drug-eluting coatings involves polymer coatings on medical devices, such as catheters and implants, releasing drugs or growth factors gradually to prevent blood clot formation, reduce inflammation, and foster tissue regeneration. In addition, antimicrobial coatings containing substances like silver or quaternary ammonium compounds are also employed to impede the growth of bacteria and microorganisms on medical devices and implants, showcasing the diverse therapeutic capabilities of drug-eluting coatings (Jose, Gizdavic-Nikolaidis, and Swift 2023).

10.2.3 Anti-adhesion coatings

Anti-adhesion coatings, designed to impede the adherence of substances on diverse substrates, exhibit key characteristics and applications that contribute to their efficacy. These coatings neutralize the surface energy and inhibit the non-specific binding to develop a hydrophilic membrane on the surface. Anti-adhesion coatings applicable to a range of entities, such as proteins, peptides, antibodies, low molecular weight substances, cells, bacteria, and parts of other living organisms, and so, the anti-adhesion coatings play a pivotal role in diverse applications (Rodichev, Gorin, and Tokmakova 2019). Notable examples include Prevelex® AP1, a nanometre-scale polymer coating preventing the adhesion of proteins, peptides, and antibodies on various substrates, and Prevelex® CC1, a similar coating preventing the cell adhesion films (Vasilev et al. 2020).

10.2.4 Hydrophilic and hydrophobic coatings

Hydrophilic and hydrophobic coatings have different applications in healthcare, particularly in medical devices and biomedical applications. Hydrophilic coatings find widespread use in various medical devices, including implants, catheters, stents, needles, and biosensors, contributing to enhanced performance and biocompatibility (Falde et al. 2016). These coatings offer notable advantages, such as reduced friction and adhesion, thereby lowering the risk of infections, inflammations, and thrombosis. In comparison, the hydrophilic coatings exhibit superior durability and stability compared to the antimicrobial and bioactive coatings, and they can be applied to various substrates like polymers, ceramics, and glass. On the other hand, the hydrophobic coatings find applications in contact lenses, drug delivery systems, and implantable medical devices, aiding in the reduction of protein fouling and subsequently minimizing bacterial and mammalian cell adhesion (Kumar and Sahani 2021). Commonly used hydrophobic materials in healthcare applications are polyethylene glycol (PEG) and polyethylene oxide (PEO). It is clear that the hydrophilic coatings primarily aim to enhance medical device performance and biocompatibility, while the hydrophobic coatings focus on reducing protein fouling and improving the durability of implantable medical devices (Hezi-Yamit et al. 2009). In addition to that, the combination of hydrophilic and hydrophobic coatings could also have an advantage in providing a balanced polymer system that facilitates the controlled drug elution with biocompatibility properties.

10.2.5 Nano coatings

Nano coatings for healthcare applications have emerged as a focal point of research, with considerable attention directed towards their potential in preventing bacterial adhesion, mitigating biofouling, and enhancing the performance of medical devices. The promising applications of nano coatings can also be evidenced in the recent literature, which are discussed as follows. Recent research on a broad-spectrum anti-adhesive coating with antimicrobial properties, derived from an extracellular polymer of a marine cyanobacterium, underscores its potential in healthcare settings (Malik, Muhammad, and Waheed 2023). Nano materials, such as nano-titanium dioxide (n-TiO_2) and nano-hydroxyapatite (n-HA), have been used in medical devices, including implants, stents, hip prostheses, and orthopaedic devices, to enhance their functionality and biocompatibility (Haleem et al. 2023). Metallic nanoparticles, such as silver nanoparticles (AgNPs), copper nanoparticles (CuNPs), and zinc oxide nanoparticles (ZnONPs), gained severe attention in nanotechnology-based antimicrobial and antiviral surface coatings to restrict the spreading of viruses and reduce bacterial and mammalian cell adhesion (Malik, Muhammad, and Waheed 2023). Carbon nanotubes (CNTs) were used in antimicrobial coatings to protect the medical instrument from contamination (Butler et al. 2023).

10.3 TYPES OF BIO-SURFACE COATINGS

The commonly used bio-surface coating in healthcare applications (Figure 10.2) is polymers, ceramics, metals, and nanomaterials, which are more detailed in the sections below.

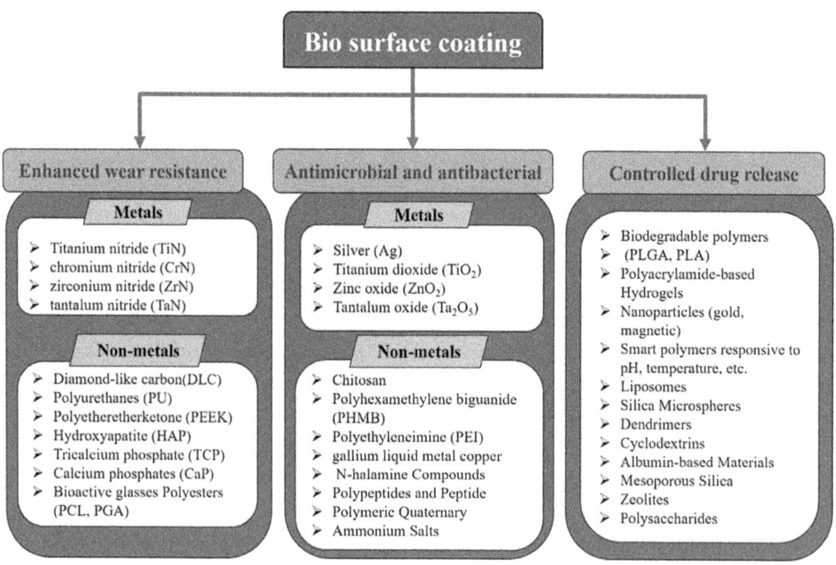

Figure 10.2 Classification of bio-material coatings in healthcare application.

10.3.1 Polymer coatings

Biopolymers are light-dense and exhibit excellent biocompatibility, making them suitable for medical applications. They are biologically renewable, biodegradable, and non-toxic, which is essential for their use in implantable medical devices and other biomedical applications (Pearce and O'Reilly 2021). Polymers, such as polystyrene (PS), parylene, polypropylene (PP), polyethylene (PE), polyurethane (PU), polyvinyl chloride (PVC), polycarbonate (PC), polyetheretherketone (PEEK), and polyethylene terephthalate (PET), are employed to coat medical devices owing to their flexibility, biocompatibility, and robustness (Maitz 2015; Li et al. 2021). Specifically, biopolymers, like parylene, are recognized for their pliability and capacity to emulate natural tissues, rendering them suitable for various medical applications. Parylene stands out as a widely utilized conformal coating for implantable medical devices, such as stents, catheters, and electrosurgical devices (Farooq et al. 2020). Table 10.1 lists the most commonly used polymer coating material and its properties. Some examples of biomedical devices that use polymer coatings are discussed in the following sections.

10.3.1.1 Cardiovascular stents

Polymer coatings, such as lactic-co-glycolic acid, tyrosine-derived polycarbonate, carbon, gold, and silicon carbide, are applied to cardiovascular stents to improve their biocompatibility and reduce the risk of restenosis. These coatings can also be used to deliver drugs to the target site, preventing the re-narrowing of the blood vessels.

10.3.1.2 Orthopaedic implants

Polymer coatings (polymethylmethacrylate (PMMA), polycaprolactone (PCL), polyethylene (PE), and ultra-high molecular weight polyethylene), along with other ceramic materials, such as hydroxyapatite, tricalcium phosphate, are utilized in orthopaedic implants to enhance their wear resistance, reduce friction, and improve their integration with the surrounding bone tissue. These combined coatings can also facilitate drug delivery to promote bone growth and prevent infection.

10.3.1.3 Catheters and guidewires

Common polymers, such as polyamide (nylon), polyether block amide, polyurethane, and thermoplastic elastomers, are employed in catheters for their biocompatibility and flexibility. Polymers are also employed in catheters and guidewires to reduce friction and improve manoeuvrability, thereby preventing the formation of blood clots. These coatings can also incorporate antimicrobial agents to reduce the risk of infections associated with

Table 10.1 Polymer coating materials and their properties

Various polymer coatings	*Properties*	*Hardness (Shore D)*	*Coating thickness (μm)*	*Coefficient of friction*
Polydimethylsiloxane (PDMS)	Biocompatible, flexible, gas permeable, optically clear	30–65	10–1000	0.2–0.4
Polyethylene (PE)	Durable, chemical-resistant, good barrier to moisture and gases	50–70	10–250	0.3–0.7
Polytetrafluoroethylene (PTFE)	Excellent low friction, non-stick, chemically inert	50–60	10–50	0.04–0.1
Polyurethane (PU)	Biocompatible, good wear resistance, tunable hardness, and elasticity	30–90	50–500	0.2–0.8
Polyvinyl chloride (PVC)	Low cost, versatile, good insulation	60–75	25–250	0.3–0.6
Hydrogel coatings	Hydrophilic, biocompatible, drug delivery capabilities	40–70	10–100	0.1–0.3
Heparin coatings	Antithrombogenic, reduces clotting on medical devices	40–50	1–10	0.2–0.4
Silicone elastomers	Biocompatible, high flexibility, shock absorbing	30–70	50–500	0.2–0.5
Fluoropolymers	Excellent chemical resistance, biocompatible (selected types)	60–70	10–100	0.2–0.5
Polycarbonate (PC)	Biocompatible, strong, transparent	65–75	25–500	0.3–0.6

these devices. Guidewires are frequently coated with fluoro paint, providing improved sliding operation. Fluoropolymer tubes are also used for the innermost layer of catheters, enhancing lubricity and reducing friction during medical procedures (Shi et al. 2020).

10.3.1.4 Implantable drug delivery systems

Polymer coatings are employed in implantable drug delivery systems to control the release of therapeutic agents. These coatings can be tailored to achieve sustained or targeted drug delivery, improving the efficacy and safety of the treatment (Magill et al. 2023). Polycaprolactone (PCL), known for its biocompatibility and biodegradability, stands out as a promising candidate for polymeric implants, providing a platform for controlled drug release (Stewart et al. 2018). Hyaluronic acid (HyA) microspheres and hydrogels have been utilized as coating substances, altering the surface of implants and enhancing osseointegration in osteogenesis applications (Hakim et al. 2021). Recent advances in materials science have introduced biodegradable polymer coatings, enabling precise and controllable drug delivery for preventing various conditions (Pan, Zhou, and Yu 2018). Implantable biomaterials, such as the Gliadel wafer comprising carmustine and a polymer, demonstrate the diversity of materials used for specific therapeutic purposes. In neurology, soft and delicate tissues require biopolymeric coatings for drug delivery in brain/neural implants, emphasizing the importance of material choice in achieving therapeutic efficacy (O. Adeosun et al. 2020).

10.3.1.5 Biosensors

Polymer coatings are essential components of biosensors, where they are used to immobilize biological recognition elements and enhance the sensor's selectivity and sensitivity. Conductive polymers play a crucial role as coatings or encapsulating materials on electrodes, contributing to the sensitivity and functionality of the biosensor (Popescu and Ungureanu 2023). Bio-coatings, encompassing a variety of natural and synthetic materials, are employed to immobilize and protect biological components within biosensors, improving their overall performance and lifespan (Huang, Zhu, and Kianfar 2021). Polymeric materials, such as chitosan, hyaluronic acid, and other biogenic nanoparticles (Song, Winkeljann, and Lieleg 2020), are used in medical surfaces to immobilize biological components, and these polymeric materials are also used in improving the stability of biosensors, aiding in the regeneration of biosensor surfaces, thereby increasing their lifespan (Popescu and Ungureanu 2023).

10.3.2 Metallic coatings

Metallic coatings have a wide range of applications in healthcare, particularly in medical devices and biomedical applications. These coatings can provide various properties, such as corrosion protection, anti-wear, antibacterial, and biocompatibility, to different substrates (Smith et al. 2022). Metal–organic framework (MOF) coatings have possible applications in surface modification for medical implants and tissue engineering. These coatings are designed to have antimicrobial properties and reduce the risk of infection. They can also act as reservoirs for controlled drug release, enabling localized treatment and improved therapy. Some of the examples of metallic coatings used in healthcare applications are discussed in the following sections.

10.3.2.1 Zinc-rich coatings

Zinc metal coatings are applied to various medical devices, such as implants and orthopaedic devices, to provide corrosion protection and antibacterial properties. They can be overcoated with other materials, such as epoxies or polyurethanes, to improve their performance and durability.

10.3.2.2 Anodic coatings

Anodic metal coatings are applied to aluminium and aluminium alloys to improve corrosion resistance and biocompatibility. The process of applying anodic coatings involves the use of electrolytic solutions and controlled current densities to create a thin, adherent oxide film on the surface of the material.

10.3.2.3 Chemical conversion coatings

Chemical conversion coatings are applied to aluminium and aluminium alloys to provide corrosion protection and improve their biocompatibility. These coatings are formed through a chemical reaction between the aluminium substrate and a conversion coating solution, resulting in the formation of a tightly adherent oxide film.

10.3.2.4 Titanium coatings

Titanium coatings, such as hydroxyapatite (HA) or titanium dioxide (TiO_2), are applied to medical devices, such as implants and stents, to improve their biocompatibility and osseointegration properties. These coatings can enhance the performance of medical devices and reduce the risk of complications or rejections.

10.3.2.5 Gold coatings

Gold coatings are used in some medical devices, such as dental restorations and electronic devices for medical applications, to provide biocompatibility, corrosion resistance, and anti-wear properties. Table 10.2 lists the most commonly used metallic coating materials and their properties.

10.3.3 Ceramic coatings

Ceramic coatings have potential applications in various healthcare fields due to their unique properties, such as biocompatibility, wear resistance, and resistance to corrosion. Ceramic coatings have several protective features for medical and dental components because of their high hardness and low coefficient of friction. Zirconia (ZrO_2) and alumina (Al_2O_3) are

Table 10.2 Metallic coating materials and their properties (Amirtharaj et al. 2022; Mohapatra and Rautray 2023)

Metallic coating	*Properties*	*Hardness (HV)*	*Coating thickness (μm)*	*Coefficient of friction*
Titanium (Ti)	Biocompatible, corrosion-resistant, strong, lightweight	300–400	5–50	0.4–0.6
Titanium nitride (TiN)	Biocompatible, wear-resistant, golden colour, enhances wear resistance	1700–2200	1–10	0.2–0.4
Chromium (Cr)	Corrosion-resistant, enhances wear resistance, hard but brittle	700–1000	0.1–10	0.4–0.7
Chromium nitride (CrN)	Corrosion-resistant, wear-resistant, high hardness, dark grey colour	1800–2500	1–10	0.2–0.4
Tantalum (Ta)	Excellent corrosion resistance, biocompatible, used in dental implants	170–280	1–10	0.4–0.6
Platinum (Pt)	Highly biocompatible, used in pacemakers and electrodes	120–150	1–5	0.3–0.5
Silver (Ag)	Antibacterial, used in catheters and wound dressings	25–35	0.1–1	0.2–0.4
Gold (Au)	Highly biocompatible, used in dental restorations and implants	35–45	0.1–1	0.3–0.5
Nitinol (NiTi)	Shape-memory alloy, used in stents and braces	300–600	10–100	0.4–0.6

the two most important ceramic oxides used for biomedical coating purposes. They can heal damaged tissue and bone joints. Ceramic coatings are harder and more wear-resistant than polymer coatings, making them ideal for use in load-bearing applications such as orthopaedic implants (Malini, Raj, and Anantharaju 2023). Ceramic coatings can withstand high temperatures, making them suitable for use in medical devices that require sterilization. Table 10.3 lists the most commonly used ceramic coating material with its properties. Ceramic materials are coated on the surfaces of metals to form bioactive layers and with antimicrobial agents to prevent infections, promote tissue regeneration and wound healing, and integrate with surrounding tissues. These coatings significantly improve the chemical stability of implants and find applications, such as load-bearing parts, joint replacements, fillers, veneering materials, drug delivery platforms, and biomimetic scaffolds.

Ceramic coatings are tailored for different applications and serve as an excellent strategy against corrosion, cytotoxicity, and chemical instability. Despite their advantages, ceramics are generally brittle, and their strength is

Table 10.3 Ceramic coating materials and their properties

Ceramic coating	*Properties*	*Hardness (HV)*	*Coating thickness (μm)*	*Coefficient of friction*
Alumina (Al_2O_3)	High wear resistance, biocompatible, chemically inert	300–400	5–50	0.4–0.6
Zirconia (ZrO_2)	High strength, fracture toughness, wear resistance, good biocompatibility	1200–1500	10–100	0.4–0.6
Hydroxyapatite (HAp)	Excellent biocompatibility, mimics bone composition, promotes bone ingrowth	400–600	50–150	0.4–0.7
Silicon nitride (Si_3N_4)	High strength, hardness, wear resistance, excellent biocompatibility	1800–2500	1–10	0.2–0.4
Titanium nitride (TiN)	High wear resistance, golden colour, enhances the wear resistance of the underlying material	1700–2200	1–10	0.2–0.4
Diamond-like carbon (DLC)	Exceptional wear resistance, low friction, biocompatible (certain types)	5000–8000	0.1–10	0.1–0.2
Tantalum oxide (Ta_2O_5)	Excellent corrosion resistance, biocompatible	700–1000	1–10	0.4–0.6
Yttria-stabilized zirconia (YSZ)	High strength, fracture toughness, good biocompatibility, thermal insulating	1000–1200	10–100	0.4–0.6
Silicon dioxide (SiO_2)	Biocompatible, good chemical resistance, optically transparent	400–600	0.1–10	0.2–0.4
Calcium phosphate (CaP)	Biocompatible, mimics bone composition, promotes bone ingrowth	400–600	50–150	0.4–0.7
Tricalcium phosphate (TCP)	Biocompatible, high strength, promotes bone resorption and remodelling	500–700	50–200	0.4–0.7
Zinc oxide (ZnO_2)	Antimicrobial, biocompatible, promotes wound healing, UV protection	400–500	0.1–10	0.3–0.5
Titanium dioxide (TiO_2)	Antimicrobial, photocatalytic, self-cleaning, biocompatible (specific forms)	300–400	0.1–10	0.4–0.6

highly dependent on the presence of flaws. To address this, nearly inert ceramic coatings can be deposited on the surface of metallic medical devices to impart wear resistance and corrosion resistance. Some common ceramic coatings used in healthcare applications include hydroxyapatite, bioactive glass, and silica-based ceramics. These coatings are used to improve the integration of implants with the surrounding bone, reduce the risk of loosening and failure, and promote cell attachment, growth, and tissue formation (Smith et al. 2022). Ceramic coatings have a wide range of potential applications in medical devices, and are discussed in the following sections.

10.3.3.1 Orthopaedic and dental implants

Ceramic coatings, such as hydroxyapatite and bioactive glass, are used to improve the integration of implants with the surrounding bone, reduce the risk of loosening and failure, and promote tissue regeneration, wound healing, and osteointegration with the surrounding tissues (Vaiani et al. 2023).

10.3.3.2 Corrosion protection

Ceramic coatings provide enhanced corrosion protection to metallic medical devices, such as orthopaedic and dental implants, ensuring their longevity and effectiveness (Smith et al. 2022).

10.4 COATING TECHNIQUES FOR HEALTHCARE APPLICATIONS

The development and improvement of advanced coating techniques have become essential in enhancing the performance, biocompatibility, and functionality of biomaterials. The coating methods play a major role in tailoring the properties of diverse biomaterials, ranging from medical implants to drug delivery systems, to meet the rigorous demands of healthcare applications. By using various advanced coating techniques, researchers and medical practitioners aim to achieve specific functionalities, such as improved adhesion, controlled drug release, and enhanced biocompatibility. The various coating techniques that are used in healthcare applications are discussed in the following sections.

10.4.1 Physical vapour deposition

Physical vapour deposition (PVD) is a coating process (Figure 10.3) used to deposit a thin film on the substrate through vapourization in a controlled manner (Geyao et al. 2020). This technique offers several advantages, including the ability to produce coatings with precise thickness, uniformity, and adhesion. PVD coatings are also highly wear-resistant, corrosion-resistant,

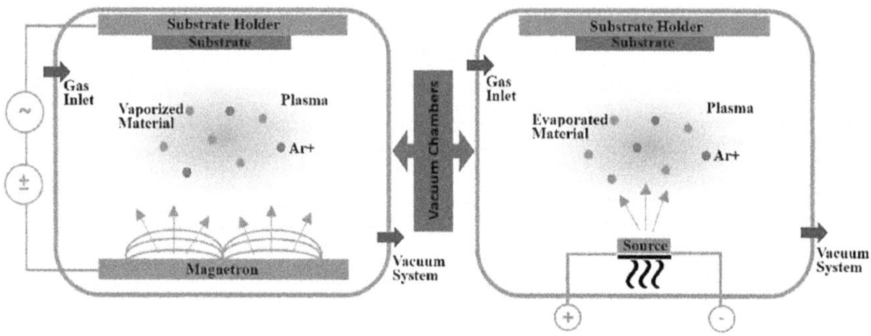

Figure 10.3 Schematic representation of PVD process.

and biocompatible, making them ideal for use in medical devices and implants (Geyao et al. 2020). Some examples of PVD coatings used in healthcare applications include zirconium nitride (ZrN), chromium nitride (CrN), and titanium nitride (TiN) coatings, which are used to improve the wear resistance and corrosion resistance of medical devices. PVD coatings are also used to produce antimicrobial coatings, lubricating coatings, and drug delivery coatings for medical devices. Zirconium nitride (ZrN) coatings produced through physical vapour deposition (PVD) exhibit excellent corrosion and wear resistance, making them well-suited for various applications, including healthcare (Franklin and Yang 2020). ZrN coatings are biocompatible, extremely tough, and highly resistant to abrasive wear, making them ideal for use in medical components and implants. These coatings have a hardness of 2800 ± 300 HV, slightly harder and smoother than titanium nitride (TiN) coatings. ZrN coatings are also biocompatible and highly resistant to abrasive wear, and they meet all requirements for biocompatibility, making them well-suited for medical components. The coatings are also highly abrasive, wear-resistant, and are suitable for applications in which coating with TiN would result in galling (Aissani et al. 2022).

However, there are also some disadvantages to PVD coatings. The process can be slow, and the coatings can be expensive to produce. In addition, the coatings may be prone to failure during heat treatment, and the process may require specialized equipment and expertise. Despite these limitations, PVD remains a valuable technology for producing high-quality coatings for medical devices and implants, contributing to their improved performance and biocompatibility in healthcare applications.

10.4.2 Chemical vapour deposition

Chemical vapour deposition (CVD) emerges as a pivotal technique in the realm of healthcare, offering a versatile and precise method for coating surfaces with thin films of various materials. This advanced deposition process involves the chemical reaction of gaseous precursors to form a solid coating

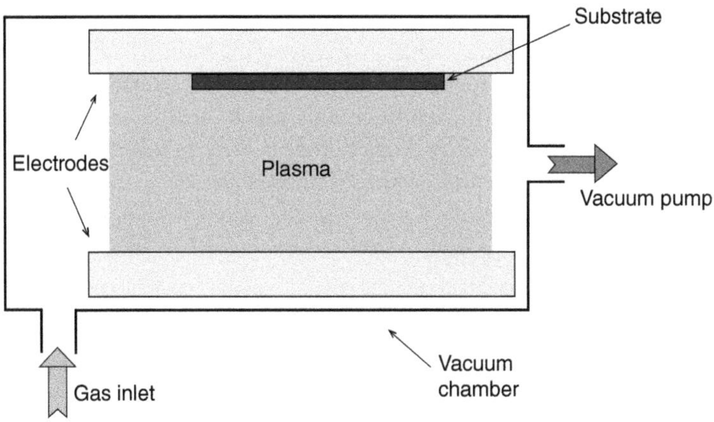

Figure 10.4 Schematic representation of CVD process.

on a substrate. Figure 10.4 illustrates the schematic representation of the CVD process. CVD relies on the controlled chemical reactions occurring in the vapour phase. During the process, precursor gases are introduced to the substrate, where they react, forming a solid thin film. This method ensures uniform coverage and precise control over the composition and thickness of the deposited coating, making CVD well-suited for healthcare applications. For example, the diamond-like carbon (DLC) coating deposited using the CVD has found applications in cardiovascular stents. The advantages are that the DLC coating deposited through CVD provides a smooth, biocompatible surface that minimizes friction, reduces the risk of thrombosis, and enhances the stent's longevity within the bloodstream.

The precise control offered by CVD ensures a uniform and adherent DLC coating, contributing to the stent's efficacy in preventing restenosis and improving overall patient outcomes. Recent advancements in CVD technology have enabled the production of coatings with specific thicknesses and uniformity, allowing for the deposition of coatings with precise properties. CVD coatings are also solvent-free and substrate-independent, making them suitable for delicate substrates and allowing control of device mechanical properties independent of the device surface chemistry. Recent research has focused on the use of tantalum (Ta) and titanium aluminide (TaAl) coatings produced through chemical vapour deposition (CVD) for bio-surface coating in healthcare applications. Tantalum coatings have been applied to medical devices and implants using CVD, offering excellent corrosion resistance and good wear resistance, making them suitable for use in orthopaedic implants and dental devices (Echavarría et al. 2017). Tantalum coatings ranging from 0.5 to 120 mm are deposited by CVD at temperatures between 625 and 1000 °C, utilizing tantalum pentachloride as a precursor. Deposition rates vary based on conditions, providing versatility in thickness and application possibilities. Recent studies have investigated the tribological behaviour of Ta-N-based coatings, which are hard, chemically inert, and corrosion-resistant, making them attractive for biomedical applications. TaAl coatings have been applied to porous Ti6Al4V scaffolds using CVD, exhibiting improved biocompatibility and mechanical properties, making them suitable for use in orthopaedic implants and dental devices (Gautam et al. 2022; Martín-Palma and Lakhtakia 2013).

10.4.3 Sol–gel processing

Sol–gel processing (Figure 10.5) is a versatile method extensively employed for bio-surface coating in healthcare applications. This technique involves the synthesis of materials from a solution or colloidal suspension that transforms into a gel-like form, facilitating the creation of thin films or coatings on various substrates. In biomedical applications, sol–gel coatings are used to modify surfaces, enhancing properties, such as bioactivity, corrosion resistance, and wear resistance. For instance, medical device designers utilize sol–gel coatings to improve the lubricity and biocompatibility of surfaces, providing functional modifications that are crucial in healthcare settings (Owens et al. 2016). The sol–gel process offers a cost-efficient approach to generate multifunctional coatings, contributing to advancements in medical devices and implants by ensuring optimal performance and biocompatibility. Researchers are actively investigating the use of metal–organic frameworks (MOFs) in sol–gel-derived coatings, showcasing their potential for enhancing surface properties in medical applications. These materials offer versatile functionalities, demonstrating promise in surface modification for medical implants and tissue engineering. In addition, there is a notable focus on superhydrophobic and antibacterial sol–gel coatings, presenting a compelling avenue for improving the biocompatibility and performance of medical devices

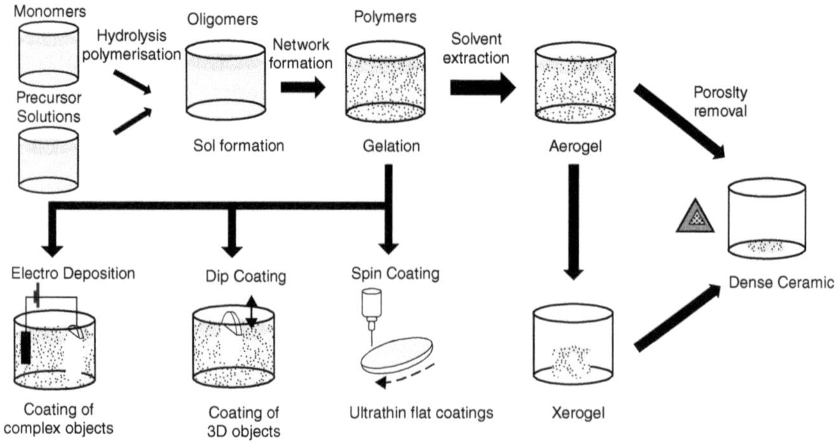

Figure 10.5 Schematic representation of sol-gel process.

(Wang et al. 2022). In recent research endeavours, the integration of hydroxyapatite (HAP) with sol–gel coatings, encompassing both polymer and ceramic matrices, has garnered significant attention for its potential applications in the biomedical field. Sol–gel-derived HAP coatings have been synthesized at relatively low temperatures, enhancing their compatibility with temperature-sensitive substrates. Investigations into the combination of HAP with polymers have shown promising results, capitalizing on the sol–gel technique's ability to produce uniform and homogenous coatings. Moreover, the incorporation of ceramic components, such as zirconia, into the sol–gel process, has been explored to enhance the biological performance of these coatings. This multifaceted approach, leveraging the versatility of sol–gel methodologies, opens avenues for tailoring the properties of HAP coatings, making them suitable for implants and tissue engineering (Zanurin et al. 2022).

10.4.4 Electrophoretic deposition

Electrophoretic deposition (EPD) emerges as a versatile and effective technique for the precise coating of biomedical surfaces in healthcare applications. In this process, the suspended charged particles in a liquid medium move towards the target substrate under the influence of an electric field. The schematic experiment used in EPD coating is displayed in Figure 10.6. EPD offers several advantages, including the ability to coat without binders, ensuring a high level of purity in the deposited material. In healthcare, EPD has found application in the creation of coatings for medical devices. For instance, hydroxyapatite (HA) coatings on implants have been achieved through EPD, showcasing excellent stoichiometry control and high purity of the deposited material. Moreover, EPD has been utilized for the construction of drug-releasing coatings on medical devices, acting as reservoirs for controlled local delivery (Batool et al. 2021).

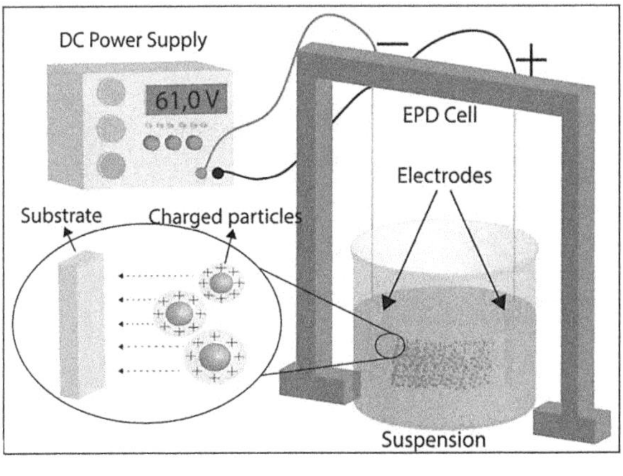

Figure 10.6 Schematic representation of EPD process.

The integration of zinc oxide (ZnO), titanium oxide (TiO_2), and hydroxyapatite (HAP) in electrophoretic coating has shown promising improvements in coating properties. Studies have investigated the deposition of zinc-doped hydroxyapatite coatings on titanium through electrophoretic deposition, highlighting enhanced hardness attributed to the controlled deposition kinetics and coating morphology achieved (Elter et al. 2019; Avcu et al. 2019). In addition, the influence of TiO_2 on the microstructure of coatings, when combined with hydroxyapatite, has revealed uniform morphology and improved mechanical properties, showcasing the potential for enhanced hardness (Dhiflaoui et al. 2023). The electrophoretic coating technique, known for its easy control over coating thickness and morphology, has demonstrated anaphoretical/oxidative approaches for in situ synthesis of coatings, offering versatility in tailoring coating properties (Pantović Pavlović et al. 2019). The surface modification of hybrid composite multilayers, including ZnO and TiO_2, has further displayed nano-scale surface morphology, advanced crystallization, and homogeneous structures, contributing to improved coating characteristics (Fahad et al. 2023). This comprehensive approach, combining ZnO, TiO2, and HAP in electrophoretic coating, holds significant promise for achieving coatings with superior hardness, controlled thickness, and enhanced morphology.

10.4.5 Plasma spraying

Plasma spraying (Figure 10.7) is a widely utilized technique in healthcare for applying bioactive coatings, particularly hydroxyapatite (HA), onto various medical implants. One notable application is in the field of alloprosthetic implants, where air plasma spraying (APS) is employed to produce HA

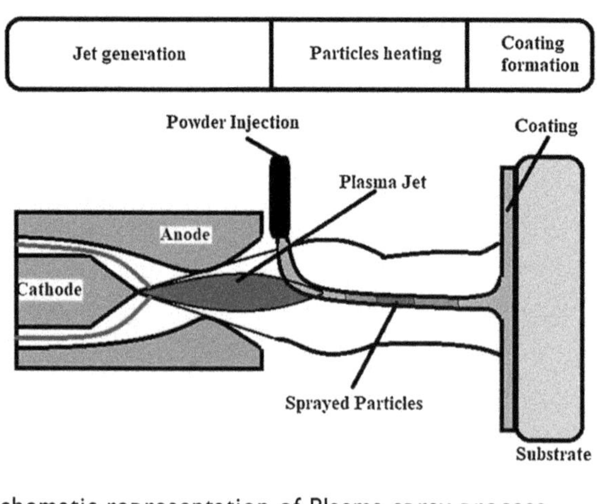

Figure 10.7 Schematic representation of Plasma spray process.

coatings. This method allows for the modification of HA spraying, enhancing its potential for improved integration with the host tissue on implant surfaces (Kowalski et al. 2022). Furthermore, a novel thermal spraying technique known as solution precursor plasma spraying (SPPS) has been employed in recent days to manufacture bioactive glass coatings. This technique offers a promising avenue for producing bioactive glass coatings with enhanced biological properties (Henao et al. 2019). The nature of plasma spraying itself involves a rapid solidification process, wherein finely powdered material is melted and solidified almost instantly in a plasma jet, resulting in coatings suitable for biomedical applications (Heimann 2023; Wang et al., 2023a).

Recent studies have focused on augmenting the properties of plasma-sprayed coatings, specifically involving Baghdadite bioceramic and Al-Ni alloys. The incorporation of carbon nanotubes in plasma-sprayed bioceramic coatings has been explored, aiming to enhance the overall properties of the coating. This investigation, which concentrated on the corrosion behaviour of plasma-sprayed Baghdadite bioceramic coatings, revealed improvements in hardness and the mechanical and biological properties of the coated substrates (Singh et al. 2024). Furthermore, studies on Al-Ni alloy coatings (healthcare instruments) produced through plasma spray have delved into the impact of manufacturing parameters on crucial coating characteristics (Wang et al. 2023b).

10.5 CHALLENGES AND FUTURE PERSPECTIVES

Bio-surface coating in healthcare applications has seen significant advancements in recent years, with various coating techniques and materials being developed for use in medical devices and implants. However, there are still challenges and future trends that need to be addressed to further improve the performance and biocompatibility of these coatings. Current challenges include a mismatch between surface chemistry and application needs, leading to discomfort or contamination in various healthcare settings. To tackle this, researchers are exploring different coating technologies. Passive surface finishing modification (PMS) is one such approach focusing on enhancing surface properties without active interventions, offering a potential solution to combat challenges related to poor surface properties. Additionally, the development of bio-functional coatings could be a competent strategy to meet diverse clinical requirements and address corrosion issues, highlighting progress in this field (Singh et al. 2023).

Antimicrobial materials face challenges in commercialization and material stability post-application, hindering their widespread adoption (de Castro and Costa 2021). Nevertheless, advancements in surface treatment for 3D-printed implants offer promising directions for overcoming current challenges, indicating the potential for bioinspired coatings to enhance implant functionality and biocompatibility (Liu et al. 2021). In the realm of metal–organic frameworks (MOFs), thin film coatings present both challenges and opportunities for biomedical applications, highlighting the need for further research to unlock their full potential (Sabzehmeidani and Kazemzad 2023). In the realm of antimicrobial surfaces, bioinspired antimicrobial peptides (AMPs) are emerging as potential components for future materials in healthcare applications (Sheridan et al. 2022). Sustainable biomaterials, including protein-based, cellulose, chitin, and chitosan composites, are also gaining attention for their properties and applications in biomedicine (Pesode et al. 2023). Moreover, the use of nanomaterials, including metals, metal oxide nanoparticles, and 2D nanomaterials, is explored for antimicrobial coatings in the biomedical field, presenting a contemporary approach to enhance performance (Sahoo et al. 2022)

Future trends may also involve the integration of smart coatings, responsive to environmental cues, further elevating the capabilities of bio-surface coatings in healthcare applications. With the rise of personalized medicine, there will be a growing demand for customized bio-surface coatings tailored to individual patient needs. Sustainability will play a key role in future research, exploring eco-friendly materials and coating techniques to minimize environmental impact while promoting biocompatibility. Furthermore, the exploration of smart materials that can dynamically respond to environmental changes, such as pH or temperature, is anticipated to be a significant area of interest in upcoming research endeavours.

10.6 CONCLUSION

In conclusion, the comprehensive exploration of bio-surface coatings in healthcare applications, as detailed in the book chapter, provides valuable insights into the evolution of this critical field. The chapter's classification of bio-surface coatings illuminates the diverse strategies employed to enhance medical devices and implants. By delving into the coating techniques utilized, the chapter highlights the significance of innovative methods in achieving optimal surface modifications. The discussion on materials used in surface coating underscores the importance of selecting appropriate biomaterials to ensure biocompatibility and functional efficacy. Bio-surface coatings stand at the forefront of innovation for enhancing healthcare applications. The importance of surface modification, focusing on biocompatibility, biointegration, and corrosion resistance, is evident in various medical contexts. This holistic examination serves as a foundation for advancing healthcare technologies, offering a roadmap for researchers and practitioners to navigate the intricate landscape of bio-surface coatings. As the field continues to evolve, the knowledge encapsulated in this chapter will undoubtedly catalyse further advancements, fostering the development of safer, more efficient, and biocompatible healthcare solutions.

REFERENCES

Aissani, Linda, Akram Alhussein, Abdul Zia, Gcina Mamba, and Sami Rtimi. 2022. "Magnetron Sputtering of Transition Metal Nitride Thin Films for Environmental Remediation." *Coatings* 12 (11): 1746. doi:10.3390/coatings12111746

Amirtharaj Mosas, Kamalan Kirubaharan, Ashok Raja Chandrasekar, Arish Dasan, Amirhossein Pakseresht, and Dušan Galusek. 2022. "Recent Advancements in Materials and Coatings for Biomedical Implants." *Gels* 8 (5): 323. doi:10.3390/gels8050323

Avcu, Egemen, Fatih E. Baştan, Hasan Z. Abdullah, Muhammad Atiq Ur Rehman, Yasemin Yıldıran Avcu, and Aldo R. Boccaccini. 2019. "Electrophoretic Deposition of Chitosan-Based Composite Coatings for Biomedical Applications: A Review." *Progress in Materials Science* 103 (June): 69–108. doi:10.1016/j.pmatsci.2019.01.001

Batool, Syeda Ammara, Abdul Wadood, Syed Wilayat Hussain, Muhammad Yasir, and Muhammad Atiq Ur Rehman. 2021. "A Brief Insight to the Electrophoretic Deposition of PEEK-, Chitosan-, Gelatin-, and Zein-Based Composite Coatings for Biomedical Applications: Recent Developments and Challenges." *Surfaces* 4 (3): 205–39. doi:10.3390/surfaces4030018

Bauer, Maria G., and Oliver Lieleg. 2023. "Bio-Macromolecular Surface Coatings for Autohesive, Transparent, Elastomeric Foils." *Macromolecular Materials and Engineering* 308 (7). doi:10.1002/mame.202200681

Butler, James, Richard D. Handy, Mathew Upton, and Alexandros Besinis. 2023. "Review of Antimicrobial Nanocoatings in Medicine and Dentistry: Mechanisms of Action, Biocompatibility Performance, Safety, and Benefits Compared to Antibiotics." *ACS Nano* 17 (8): 7064–92. doi:10.1021/acsnano.2c12488

Castro, Karine Cappuccio de, and Josiel Martins Costa. 2021. "Polymeric Surfaces with Biocidal Action: Challenges Imposed by the SARS-CoV-2, Technologies Employed, and Future Perspectives." *Journal of Polymer Research* 28 (6): 230. doi:10.1007/s10965-021-02548-4

Choi, Yonghyun, Binh Phan, Masayoshi Tanaka, Jinkee Hong, and Jonghoon Choi. 2020. "Methods and Applications of Biomolecular Surface Coatings on Individual Cells." *ACS Applied Bio Materials* 3 (10): 6556–70. doi:10.1021/acsabm.0c00867

Dhiflaoui, Hafedh, Sarra Ben Salem, Mohamed Salah, Youssef Dabaki, Slah Chayoukhi, Bilel Gassoumi, Anouar Hajjaji, Ahmed Ben Cheikh Larbi, Mosbah Amlouk, and Hicham Benhayoune. 2023. "Influence of TiO2 on the Microstructure, Mechanical Properties and Corrosion Resistance of Hydroxyapatite HaP + TiO2 Nanocomposites Deposited Using Spray Pyrolysis." *Coatings* 13 (7): 1283. doi:10.3390/coatings13071283

Echavarría, Aida M., P. Rico, J.L. Gómez Ribelles, Miguel A. Pacha-Olivenza, María-Coronada Fernández-Calderón, and Gilberto Bejarano-G. 2017. "Development of a Ta/TaN/TaNx(Ag)y/TaN Nanocomposite Coating System and Bio-Response Study for Biomedical Applications." *Vacuum* 145 (November): 55–67. doi:10.1016/j.vacuum.2017.08.020

Elter, Patrick, Yunis Alnazal Alramadan, Claus Moseke, and Elke Vorndran. 2019. "Electrophoretic Deposition of Zinc-Doped Hydroxyapatite Coatings on Titanium: Deposition Kinetics and Coating Morphology." *International Journal of Surface Science and Engineering* 13 (2/3): 201. doi:10.1504/IJSURFSE.2019.10024018

Fahad, Nesreen Dakhel, Nabaa Sattar Radhi, Zainab S. Al-Khafaji, and Abass Ali Diwan. 2023. "Surface Modification of Hybrid Composite Multilayers Spin Cold Spraying for Biomedical Duplex Stainless Steel." *Heliyon* 9 (3): e14103. doi:10.1016/j.heliyon.2023.e14103

Falde, Eric J., Stefan T. Yohe, Yolonda L. Colson, and Mark W. Grinstaff. 2016. "Superhydrophobic Materials for Biomedical Applications." *Biomaterials* 104 (October): 87–103. doi:10.1016/j.biomaterials.2016.06.050

Farooq, Tahir, Arruje Hameed, Muhammad Sajid Hamid Akash, and Kanwal Rehman. 2020. "Polymer Coatings for Biomedical Applications." In *Polymer Coatings*, 333–46. Wiley. doi:10.1002/9781119655145.ch14

Franklin, Trevor, and Rong Yang. 2020. "Vapor-Deposited Biointerfaces and Bacteria: An Evolving Conversation." *ACS Biomaterials Science & Engineering* 6 (1): 182–97. doi:10.1021/acsbiomaterials.9b01496

Gautam, Sanjeev, Dhruv Bhatnagar, Deepika Bansal, Hemant Batra, and Navdeep Goyal. 2022. "Recent Advancements in Nanomaterials for Biomedical Implants." *Biomedical Engineering Advances* 3 (June): 100029. doi:10.1016/j.bea.2022.100029

Geyao, Lan, Deng Yang, Chen Wanglin, and Wang Chengyong. 2020. "Development and Application of Physical Vapor Deposited Coatings for Medical Devices: A Review." *Procedia CIRP* 89: 250–62. doi:10.1016/j.procir.2020.05.149

Hakim, Lotfollah Kamali, Mohsen Yazdanian, Mostafa Alam, Kamyar Abbasi, Hamid Tebyaniyan, Elahe Tahmasebi, Danial Khayatan, Alexander Seifalian, Reza Ranjbar, and Alireza Yazdanian. 2021. "Biocompatible and Biomaterials Application in Drug Delivery System in Oral Cavity." *Evidence-based Complementary and Alternative Medicine* 2021 (November): 1–12. doi:10.1155/2021/9011226

Haleem, Abid, Mohd Javaid, Ravi Pratap Singh, Shanay Rab, and Rajiv Suman. 2023. "Applications of Nanotechnology in Medical Field: A Brief Review." *Global Health Journal* 7 (2): 70–77. doi:10.1016/j.glohj.2023.02.008

Heimann, Robert B. 2023. "The Nature of Plasma Spraying." *Coatings* 13 (3): 622. doi:10.3390/coatings13030622

Henao, John, Carlos Poblano-Salas, Mónica Monsalve, Jorge Corona-Castuera, and Oscar Barceinas-Sanchez. 2019. "Bio-Active Glass Coatings Manufactured by Thermal Spray: A Status Report." *Journal of Materials Research and Technology* 8 (5): 4965–84. doi:10.1016/j.jmrt.2019.07.011

Hezi-Yamit, Ayala, Carol Sullivan, Jennifer Wong, Laura David, Mingfei Chen, Peiwen Cheng, David Shumaker, Josiah N. Wilcox, and Kishore Udipi. 2009. "Impact of Polymer Hydrophilicity on Biocompatibility: Implication for DES Polymer Design." *Journal of Biomedical Materials Research Part A* 90A (1): 133–41. doi:10.1002/jbm.a.32057

Huang, Xiaoping, Yufang Zhu, and Ehsan Kianfar. 2021. "Nano Biosensors: Properties, Applications and Electrochemical Techniques." *Journal of Materials Research and Technology* 12 (May): 1649–72. doi:10.1016/j.jmrt.2021.03.048

Jose, Ajay, Marija Gizdavic-Nikolaidis, and Simon Swift. 2023. "Antimicrobial Coatings: Reviewing Options for Healthcare Applications." *Applied Microbiology* 3 (1): 145–74. doi:10.3390/applmicrobiol3010012

Kışla, Duygu, Gökhan Gurur Gökmen, Gulsun Akdemir Evrendilek, Tamer Akan, Tomáš Vlčko, Piotr Kulawik, Anet Režek Jambrak, and Fatih Ozogul. 2023. "Recent Developments in Antimicrobial Surface Coatings: Various Deposition Techniques with Nanosized Particles, Their Application and Environmental Concerns." *Trends in Food Science & Technology* 135 (May): 144–72. doi:10.1016/j.tifs.2023.03.019

Kowalski, Szymon, Weronika Gonciarz, Radosław Belka, Anna Góral, Magdalena Chmiela, Łukasz Lechowicz, Wiesław Kaca, and Wojciech Żórawski. 2022. "Plasma-Sprayed Hydroxyapatite Coatings and Their Biological Properties." *Coatings* 12 (9): 1317. doi:10.3390/coatings12091317

Kumar, Ravinder, and Ashish Kumar Sahani. 2021. "Role of Superhydrophobic Coatings in Biomedical Applications." *Materials Today: Proceedings* 45: 5655–59. doi:10.1016/j.matpr.2021.02.457

Li, He-yang, Dan-ni Huang, Ke-feng Ren, and Jian Ji. 2021. "Inorganic-Polymer Composite Coatings for Biomedical Devices." *Smart Materials in Medicine* 2: 1–14. doi:10.1016/j.smaim.2020.10.002

Liu, Junyi, Nafisah Bte Mohd Rafiq, Lai Mun Wong, and Shijie Wang. 2021. "Surface Treatment and Bioinspired Coating for 3D-Printed Implants." *Frontiers in Chemistry* 9 (November). doi:10.3389/fchem.2021.768007

Magill, Elizabeth, Sara Demartis, Elisabetta Gavini, Andi Dian Permana, Raghu Raj Singh Thakur, Muhammad Faris Adrianto, David Waite, et al. 2023. "Solid Implantable Devices for Sustained Drug Delivery." *Advanced Drug Delivery Reviews* 199 (August): 114950. doi:10.1016/j.addr.2023.114950

Maitz, M.F. 2015. "Applications of Synthetic Polymers in Clinical Medicine." *Biosurface and Biotribology* 1 (3): 161–76. doi:10.1016/j.bsbt.2015.08.002

Malik, Shiza, Khalid Muhammad, and Yasir Waheed. 2023. "Emerging Applications of Nanotechnology in Healthcare and Medicine." *Molecules* 28 (18): 6624. doi:10.3390/molecules28186624

Malini, S., K. Raj, and K.S. Anantharaju. 2023. "Role of Ceramics and Ceramic Coatings in Biomedical Applications." In *Advanced Ceramic Coatings for Biomedical Applications*, 1–13. Elsevier. doi:10.1016/B978-0-323-99626-6.00013-5

Martín-Palma, Raúl J., and Akhlesh Lakhtakia. 2013. "Vapor-Deposition Techniques." In *Engineered Biomimicry*, 383–98. Elsevier. doi:10.1016/B978-0-12-415995-2.00015-5

Mohapatra, Bijayinee, and Tapash R. Rautray. 2023. "Ceramic Coatings for Wound Healing Applications." In *Advanced Ceramic Coatings for Biomedical Applications*, 269–92. Elsevier. doi:10.1016/B978-0-323-99626-6.00004-4

Mubeen, Bismillah, Aunza Nayab Ansar, Rabia Rasool, Inam Ullah, Syed Sarim Imam, Sultan Alshehri, Mohammed M. Ghoneim, Sami I. Alzarea, Muhammad Shahid Nadeem, and Imran Kazmi. 2021. "Nanotechnology as a Novel Approach in Combating Microbes Providing an Alternative to Antibiotics." *Antibiotics* 10 (12): 1473. doi:10.3390/antibiotics10121473

O. Adeosun, Samson, Margaret O. Ilomuanya, Oluwashina P. Gbenebor, Modupeola O. Dada, and Cletus C. Odili. 2020. "Biomaterials for Drug Delivery: Sources, Classification, Synthesis, Processing, and Applications." In *Advanced Functional Materials*. IntechOpen. doi:10.5772/intechopen.93368

Owens, Gareth J., Rajendra K. Singh, Farzad Foroutan, Mustafa Alqaysi, Cheol-Min Han, Chinmaya Mahapatra, Hae-Won Kim, and Jonathan C. Knowles. 2016. "Sol–Gel Based Materials for Biomedical Applications." *Progress in Materials Science* 77 (April): 1–79. doi:10.1016/j.pmatsci.2015.12.001

Pan, Chenhao, Zubin Zhou, and Xiaowei Yu. 2018. "Coatings as the Useful Drug Delivery System for the Prevention of Implant-Related Infections." *Journal of Orthopaedic Surgery and Research* 13 (1): 220. doi:10.1186/s13018-018-0930-y

Pantović Pavlović, Marijana R., Sanja G. Eraković, Miroslav M. Pavlović, Jasmina S. Stevanović, Vladimir V. Panić, and Nenad L. Ignjatović. 2019. "Anaphoretical/

Oxidative Approach to the in-Situ Synthesis of Adherent Hydroxyapatite/Titanium Oxide Composite Coatings on Titanium." *Surface and Coatings Technology* 358 (January): 688–94. doi:10.1016/j.surfcoat.2018.12.003

Pearce, Amanda K., and Rachel K. O'Reilly. 2021. "Polymers for Biomedical Applications: The Importance of Hydrophobicity in Directing Biological Interactions and Application Efficacy." *Biomacromolecules* 22 (11): 4459–69. doi:10.1021/acs.biomac.1c00434

Pesode, Pralhad, Shivprakash Barve, Sagar V. Wankhede, and Akbar Ahmad. 2023. "Sustainable Materials and Technologies for Biomedical Applications." *Advances in Materials Science and Engineering* 2023 (November): 1–22. doi:10.1155/2023/6682892

Pietsch, F., A.J. O'Neill, A. Ivask, H. Jenssen, J. Inkinen, A. Kahru, M. Ahonen, and F. Schreiber. 2020. "Selection of Resistance by Antimicrobial Coatings in the Healthcare Setting." *Journal of Hospital Infection* 106 (1): 115–25. doi:10.1016/j.jhin.2020.06.006

Popescu, Melania, and Camelia Ungureanu. 2023. "Biosensors in Food and Healthcare Industries: Bio-Coatings Based on Biogenic Nanoparticles and Biopolymers." *Coatings* 13 (3): 486. doi:10.3390/coatings13030486

Rodichev, Alexey, Andrey Gorin, and Maria Tokmakova. 2019. "Adhesive Strength Research of Film Antifriction Coatings." *Materials Today: Proceedings* 19: 2329–32. doi:10.1016/j.matpr.2019.07.681

Sabzehmeidani, Mohammad Mehdi, and Mahmood Kazemzad. 2023. "Recent Advances in Surface-Mounted Metal–Organic Framework Thin Film Coatings for Biomaterials and Medical Applications: A Review." *Biomaterials Research* 27 (1): 115. doi:10.1186/s40824-023-00454-y

Sahoo, Jyotirmayee, Sanchita Sarkhel, Nivedita Mukherjee, and Amit Jaiswal. 2022. "Nanomaterial-Based Antimicrobial Coating for Biomedical Implants: New Age Solution for Biofilm-Associated Infections." *ACS Omega* 7 (50): 45962–80. doi:10.1021/acsomega.2c06211

Sheridan, Mark, Caitriona Winters, Fernanda Zamboni, and Maurice N. Collins. 2022. "Biomaterials: Antimicrobial Surfaces in Biomedical Engineering and Healthcare." *Current Opinion in Biomedical Engineering* 22 (June): 100373. doi:10.1016/j.cobme.2022.100373

Shi, Huiyu, Jonathan Wang, Katherine Vorvolakos, Kathleen White, and Nandini Duraiswamy. 2020. "Pre-Clinical Evaluation of Surface Coating Performance in Guidewire Surrogates: Potential Implications for Coated Interventional Surgical Devices." *Journal of Biomaterials Applications* 34 (7): 928–41. doi:10.1177/0885328219884453

Singh, Navdeep, Uma Batra, Kamal Kumar, Neeraj Ahuja, and Anil Mahapatro. 2023. "Progress in Bioactive Surface Coatings on Biodegradable Mg Alloys: A Critical Review towards Clinical Translation." *Bioactive Materials* 19 (January): 717–57. doi:10.1016/j.bioactmat.2022.05.009

Singh, Saminderpreet, Avneesh Kumar, Mohit Kamboj, Bodhisatwa Das, Khushboo Rakha, and Harpreet Singh. 2024. "Corrosion Behaviour of Plasma-Sprayed Baghdadite Bioceramic Coatings Reinforced with Carbon Nanotubes." *Journal of Alloys and Compounds* 976 (March): 173094. doi:10.1016/j.jallcom.2023.173094

Smith, J. R., D. A. Lamprou, C. Larson, and S. J. Upson. 2022. "Biomedical Applications of Polymer and Ceramic Coatings: A Review of Recent Developments." *Transactions of the IMF* 100 (1): 25–35. doi:10.1080/00202967.2021.2004744

Song, Jian, Benjamin Winkeljann, and Oliver Lieleg. 2020. "Biopolymer-Based Coatings: Promising Strategies to Improve the Biocompatibility and Functionality of Materials Used in Biomedical Engineering." *Advanced Materials Interfaces* 7 (17). doi:10.1002/admi.202000850

Stewart, Sarah, Juan Domínguez-Robles, Ryan Donnelly, and Eneko Larrañeta. 2018. "Implantable Polymeric Drug Delivery Devices: Classification, Manufacture, Materials, and Clinical Applications." *Polymers* 10 (12): 1379. doi:10.3390/polym10121379

Vaiani, Lorenzo, Antonio Boccaccio, Antonio Emmanuele Uva, Gianfranco Palumbo, Antonio Piccininni, Pasquale Guglielmi, Stefania Cantore, Luigi Santacroce, Ioannis Alexandros Charitos, and Andrea Ballini. 2023. "Ceramic Materials for Biomedical Applications: An Overview on Properties and Fabrication Processes." *Journal of Functional Biomaterials* 14 (3): 146. doi:10.3390/jfb14030146

Vasilev, Semen, Andrey Vodyashkin, Daria Vasileva, Pavel Zelenovskiy, Dmitry Chezganov, Vladimir Yuzhakov, Vladimir Shur, Emmet O'Reilly, and Alexandr Vinogradov. 2020. "An Investigative Study on the Effect of Pre-Coating Polymer Solutions on the Fabrication of Low Cost Anti-Adhesive Release Paper." *Nanomaterials* 10 (8): 1436. doi:10.3390/nano10081436

Wang, Huanhuan, Naiming Lin, Meisam Nouri, Zhiqi Liu, Yuan Yu, Qunfeng Zeng, Guanshui Ma, Jianfeng Fan, Dongyang Li, and Yucheng Wu. 2023a. "Improvement in Surface Performance of Stainless Steel by Nitride and Carbon-Based Coatings Prepared via Physical Vapor Deposition for Marine Application." *Journal of Materials Research and Technology* 27 (November): 6021–46. doi:10.1016/j.jmrt.2023.10.307

Wang, Leijie, Xudong Guo, Hongmei Zhang, Yinxia Liu, Yongxin Wang, Kun Liu, Haofang Liang, and Wuyi Ming. 2022. "Recent Advances in Superhydrophobic and Antibacterial Coatings for Biomedical Materials." *Coatings* 12 (10): 1469. doi:10.3390/coatings12101469

Wang, Shenglian, Shuang Chen, Ming Liu, Qinghua Huang, Zimo Liu, Xin Li, and Shaofeng Xu. 2023b. "The Effect of Plasma Spray Parameters on the Quality of Al-Ni Coatings." *Coatings* 13 (12): 2063. doi:10.3390/coatings13122063

Zanurin, A., N.A. Johari, J. Alias, H. Mas Ayu, N. Redzuan, and S. Izman. 2022. "Research Progress of Sol-Gel Ceramic Coating: A Review." *Materials Today: Proceedings* 48: 1849–54. doi:10.1016/j.matpr.2021.09.203

Chapter 11

Solid lubricants

Enhancing reliability in extreme environments

Sheikh Haris Mukhtar, Chandra Shekhar, M. F. Wani, Jagtar Singh, and Umida Ziyamukhamedova

11.1 INTRODUCTION

Solid lubricants (SLs), a special class of lubricants, are applied as a powder or thin film to reduce friction and wear on surfaces that move against one another. SLs are particularly useful in situations where liquid lubricants are impractical or ineffective, such as in vacuums, space technology, and automotive applications [1, 2]. Unlike liquid lubricants, SLs rely on the physical properties of their particles, such as shape, size, and crystal structure, to minimize friction. Their primary function is to form a continuous, adherent layer on rubbing surfaces, either mechanically, (electro) chemically, or physically. SLs excel in extreme conditions where liquid lubricants fail to provide adequate separation between bearing surfaces. SLs are particularly valuable in applications where high loads are applied to sliding surfaces in boundary or mixed lubrication regimes, at low hydrodynamically effective speeds, or over a wide temperature range. In addition, SLs are essential in high-vacuum applications, nuclear reactors, and situations where contamination from liquid lubricants is unacceptable.

SLs encompass a wide range of materials with distinct lubricating mechanisms. These lubricants can be broadly categorized into structural lubricants, reactive soaps, and mechanical and chemically active lubricants. Structural lubricants, such as graphite, rely on their layered lattice structures to provide lubrication. Mechanical lubricants, on the other hand, encompass self-lubricating organic compounds, naturally occurring metal oxide films, chemical surface coatings, and glasses. Reactive soaps, typically salts of fatty acids, are commonly used in conjunction with zinc phosphate coatings for solid metal-forming lubrication. Soaps, also known as metal salts of fatty acids, play a crucial role in the preparation of greases and can be formed in situ on metal surfaces through chemical reactions. Chemically active lubricants, including extreme pressure (EP) and anti-wear (AW) additives, interact with metal surfaces to form lubricating or protective layers. Important properties of solid lubricants include crystal structure, thermal and oxidation stability, volatility, chemical reactivity, melting point, and hardness. Heat stability, closely linked to chemical stability, is essential to prevent the

 DOI: 10.1201/9781003717362-11

thermal decomposition of the solid lubricant. Solid lubricants generally operate within a specific temperature range. For instance, graphite can effectively function at temperatures up to 650 °C under moderate loads [3].

SLs are commonly found in powder form, integrated into a resin or wax film, or as dispersions suspended in either oil or water. For powdered solid lubricants to effectively form a lubricating film, they must have suitable properties like cohesion and adhesion. Liquid lubricants like dispersions and suspensions in water or oil are used to coat surfaces for both cold and hot forming processes. In addition, oil-based dispersions and suspensions serve as additives in gear and lubrication systems. Greases can contain solid lubricating materials, such as graphite, molybdenum disulfide, zinc sulfide, boron nitride, and polytetrafluoroethylene. They can hold relatively high concentrations of these solids. For most uses, an average particle size of 5–15 μm is considered acceptable for micronized solids. Incorporating solid lubricants into greases boosts their pressure absorption ability and resistance to wear and friction. Pastes are essentially solid lubricants suspended in a carrier oil.

A common characteristic of dry lubricants is their solid state or, at least, a waxy consistency at ambient temperatures. To facilitate application to metal surfaces, the lubricant should exhibit a relatively low viscosity. To overcome the challenge of applying solids or waxes, they are often diluted before use, commonly with aqueous dispersions of wax or other solid organic compounds. Dry-film lubricants, containing solid lubricants such as MoS_2, graphite, PTFE, or starch, suspended in a solution of inorganic or organic binding agents, are applied using methods such as dipping or spraying. Organic resins, cellulose, and inorganic silicates and phosphates are the primary types of binding agents employed. Hydrocarbons and water serve as solvents, while emulsifiers aid in dispersing the product in water. Dry-film lubricants can be applied using various methods, including dipping and spraying. Upon evaporation of the solvent, often water, a wax or polymer film is formed. This process can be accelerated through heating. Once hardened, lubricating varnishes produce a highly adhesive, dry film of lubricant.

Numerous reviews extensively cover various lubricants utilized in diverse industrial applications. However, there is a notable gap in the literature concerning reviews dedicated to the exclusive use of self-lubricating materials in extreme conditions. This comprehensive review aims to fill that void by offering a detailed examination of the friction and wear behavior exhibited by self-lubricating materials in challenging environments. The study delves into the dispersion of solid lubricants within different matrices, including metals and intermetallics. Furthermore, it meticulously explores the lubrication mechanisms of self-lubricating coatings and composites, specifically under diverse challenging conditions. The review concludes by shedding light on the varied applications and associated challenges of self-lubricating materials, providing a thorough understanding of their potential in addressing demands within demanding industrial scenarios.

11.2 DIVERSE FORMS OF SOLID LUBRICANTS

SLs are classified into two main types: organic and inorganic. Organic SLs are mainly polymer-based and operate between cryogenic temperatures and near 200 °C. Inorganic SLs, on the other hand, are capable of providing good lubrication around 800 °C under certain conditions. Various inorganic solid lubricants, such as graphite, CaF_2, hBN, and MoS_2, are recognized for delivering superior lubrication in diverse environmental conditions across various applications. Inorganic solid lubricants are further divided into layered structure solid lubricants, soft metals solid lubricants, oxides of solid lubricants, and fluoride-based solid lubricants. Layered structure solid lubricants have a combination of closely packed layers, where the atoms are bound tightly together. However, the bonds between the two layers are not strong enough, which allows the layers to slide over each other and provide lubrication. Soft metals and solid lubricants are used as solid lubricants due to their low shear strength. However, they can create microstructural defects during sliding, which can limit their effectiveness at high temperatures. Oxides of SLs have high-temperature stability, but they have poor adhesion properties. Fluorides are the most preferred high-temperature SLs due to their onset of softening effect around 500 °C. However, they have poor tribological behavior at room temperature due to the severe abrasion phenomenon. Table 11.1 shows the different types of solid lubricants.

11.3 INFLUENCE OF SOLID LUBRICANTS ON TRIBOLOGICAL CHARACTERISTICS

Solid lubricants, their mechanisms, temperature limits, and friction coefficients are summarized in Table 11.2.

Metals such as silver, gold, copper, and bismuth provide lubricating properties through their high ductility and tendency for plastic deformation under shear stresses [4, 29]. Their weak interatomic bonding allows atomic planes to slip over one another, accommodating frictional loads and providing low shear strength. Silver, in particular, exhibits good thermal conductivity (430 W/mK) and has been frequently studied as a solid lubricant additive [30]. At temperatures below 400 °C, composites and coatings with soft metals, such as Ag, Au, and Cu, demonstrate significantly reduced friction and wear compared to unlubricated surfaces [5, 7, 31–33]. The metals deform plastically to generate a tribofilm and prevent direct contact of asperities. However, soft metals tend to exhibit substantial material transfer and adhesion above 400 °C, which accelerates wear [6, 34]. The increased tendency for extrusion and expulsion limits their effectiveness at higher temperatures. Overall, soft metals provide lubricity through plastic deformation up to 400 °C. But their softening at elevated temperatures requires composites optimized for mechanical strength, along with synergies with high-temperature lubricants.

Table 11.1 The different types of solid lubricants

Type of solid lubricant	*Description*	*Advantages*	*Disadvantages*
Organic SLs	Mainly polymer-based	Operate between cryogenic temperatures and near 200 °C	Not effective at high temperatures
Inorganic SLs	Capable of providing good lubrication around 800 °C under certain conditions	More effective at high temperatures	More expensive and difficult to process
Layered structure SLs	Have a combination of closely packed layers, where the atoms are bound tightly together	Provide good lubrication at high temperatures	Difficult to synthesize
Soft metals SLs	Have low shear strength	Provide good lubrication at low temperatures	Develop microstructural defects during sliding
Oxides of SLs	Have high-temperature stability	Difficult to adhere to surfaces	—
Fluorides-based SLs	Have the onset of softening effect around 500 °C	Difficult to synthesize	—

Table 11.2 Summary of solid lubricant classes

Lubricant class	*Mechanism*	*Effective temperature range*	*Typical CoF*	*References*
Soft metals	Plastic deformation	Up to 400 °C	0.2–0.6	[4–7]
Laminar solids	Easy shear of layers	MoS_2 – 350 °C Graphite – 500 °C	0.080.10–0.25	[5, 8–14]
Alkaline earth fluorides	Ductility above 500 °C	500–900 °C	0.2–0.4	[15–22]
Lubricious oxides	Easy shear of Magneli phases	Up to 1000 °C	0.1–0.3	[23–28]

Graphite is a layered allotrope of carbon with a hexagonal crystal structure. The weak van der Waals forces between the layers allow them to shear easily, providing graphite's lubricating properties [35]. It provides effective lubrication up to approximately 500 °C in environments with water vapor or oxygen. However, graphite begins to oxidize to CO and CO_2 at temperatures above 400 °C, which severely diminishes its lubricating performance [11, 36]. To extend the working temperature range, research has focused on coatings that provide an oxidation barrier, such as SiC, $MoSi_2$, or oxides/carbides/nitrides of transition metals, like W, Mo, and Ti [37]. These coatings hinder the diffusion of oxygen to the graphite. Composite coatings with graphite have exhibited lower friction and wear up to 600 °C [8, 10, 38]. But a detailed understanding of the synergistic wear mechanisms between graphite and other solid lubricants at high temperatures is still lacking. Overall, graphite provides good lubrication through interlayer shearing up to 500 °C, but oxidation resistance must be improved via protective coatings or composites to broaden its working temperature range.

Graphene is a single atomic layer of graphite with remarkable properties, including extremely high thermal conductivity and strength [39, 40]. Like graphite, graphene exhibits easy shearing between its atomic layers that can provide low friction and wear. Research has shown that graphene and few-layer graphene can reduce friction and wear for steel interfaces by three to four orders of magnitude compared to unlubricated surfaces [41]. When incorporated into metal matrix composites, graphene has exhibited low friction up to 400–550 °C [42, 43]. However, oxidation and loss of integrity for graphene-based lubricating films start to occur between 550 and 600 °C. This oxidation leads to higher, unstable friction and wear due to the loss of graphene's lubricating layer [44]. Outstanding issues for the use of graphene as a solid lubricant include better understanding the mechanisms of wear protection and interaction with mating surfaces during sliding. Most research has focused on friction and wear testing, while atomic-level analysis of the tribological behavior of graphene is still needed. In addition, more work is required on synergistic effects between graphene and other solid lubricants to extend its working temperature range. While holding great potential, graphene has not yet demonstrated a clear advantage over other solid lubricants, such as graphite and MoS_2, in terms of temperature range and friction/wear performance.

Hexagonal boron nitride (hBN) has a layered crystal structure similar to graphite. In theory, this should allow easy interlayer sliding and lubricating properties. hBN also has excellent thermal conductivity and chemical/oxidation resistance up to 1000 °C [45, 46]. However, conflicting results have been reported on the effectiveness of hBN as a solid lubricant. Some studies showed friction and wear reduction for hBN-based coatings and composites from room temperature up to 800 °C. But others found poor tribological performance, with increasing friction and wear as the temperature rose above 300–400 °C [47–49]. Issues, such as coating quality, adhesion, and optimizing hBN content, likely impact the performance. Suggested optimal hBN content ranges from 5 to 15 vol% [50–52]. But, the control of grain size, uniformity, and bonding within coatings remains difficult.

Molybdenum disulfide (MoS_2) possesses a layered hexagonal crystal structure, with weak van der Waals bonding between S–Mo–S layers. This allows the atomic layers to slide easily over one another, providing low interlayer friction with a coefficient around 0.08 [14, 53]. In inert environments, MoS_2 provides effective lubrication up to approximately 1000 °C. However, it begins to oxidize to MoO_3 above 350 °C in air, which negates its lubricating properties [54–56]. Some studies have shown that adding MoS_2 to metal or ceramic matrices can reduce friction and wear up to 700–800 °C [12, 57]. The synergistic effects with the matrix composition allow the formation of lubricating tribofilms. But the performance varies based on factors such as MoS_2 content, environmental humidity, and synergies with other solid lubricants. Optimization is needed to balance lubricant concentration with adequate mechanical properties. Overall, MoS_2 offers proven solid lubrication through interlayer shearing, but its moderate oxidation resistance limits high-temperature effectiveness.

Calcium fluoride (CaF_2) is an ionic solid lubricant that transitions from brittle to ductile behavior above 500 °C. The ductility provides lubrication between sliding surfaces up to approximately 900 °C. However, CaF_2 demonstrates poor lubricating properties at lower temperatures, where its wear behavior is abrasive [58–60]. Research has focused on composite approaches, where CaF_2 shows synergistic effects with other lubricants, such as silver, graphite, and MoS_2 [61–63]. The composites provide low friction and wear through transfer film formation from room temperature up to 600 °C. At higher temperatures from 700 to 900 °C, CaF_2 reacts with the matrix to form lubricating tribofilms [20, 64]. But CaF_2 phase content and uniformity must be balanced to avoid mechanical property degradation. Overall, CaF_2 on its own is ineffective at low-moderate temperatures but becomes a good high-temperature lubricant above 500 °C. Composites with soft metals, graphite, and transition metal dichalcogenides have shown promising synergistic lubrication from room temperature to 900 °C.

Several transition metal oxides provide solid lubrication through easy shearing mechanisms enabled by their crystal structures. Oxides of molybdenum, tungsten, vanadium, and titanium have been shown to form Magnéli phases at high temperatures above 500 °C [65, 66]. These thermodynamically stable stoichiometric structures contain shear planes with a relatively low density of metal–oxygen bonds [23]. This provides easy glide between atomic layers and low shear strength. Magnéli phases of WO_3, MoO_3, TiO_2, and V_2O_5 have exhibited low friction from 500 °C to over 1000 °C [24, 26, 67, 68]. The layered crystal structure provides a self-lubricating mechanism without material transfer or plastic deformation. Challenges include preventing oxidation at lower temperatures, where wear is often severe. Nanocomposite coatings and synergies with laminar solids can extend effectiveness down to 200–300 °C [69, 70]. Overall, Magnéli phases of certain transition metal oxides impart solid lubrication at ultra-high temperatures above 500 °C. Their layered lattice enables easy shear without material transfer. Composites and synergies with metal oxides or fluorides have shown promise to provide lubricating films up to 800 °C. Table 11.3 summarizes the wear rates reported for some representative composites containing different solid lubricants across a range of temperatures. The

Table 11.3 Wear rates of composites with solid lubricants

Matrix material	*Solid lubricant*	*Wear rate (mm^3/Nm)*	*Temperature (°C)*	*Reference*
NiCr alloy + Mo	Silver	2.0×10^{-4}	400	[32]
TiAl alloy	Silver	1.0×10^{-5}	400	[6]
NiAl alloy	MoS_2	1.0×10^{-5}	800	[12]
$ZrO_2 + Y_2O_3$	MoS_2	2.0×10^{-4}	1000	[13]
Ni_3Si alloy	MoS_2	5.0×10^{-5}	400	[71]
NiCr alloy	hBN	1.0×10^{-3}	600	[51]
Ni alloy	hBN	2.0×10^{-5}	600	[50]
Ni_3Al alloy	WS_2	2.0×10^{-5}	800	[72]
TiAl alloy	WS_2	1.0×10^{-4}	600	[73]
Al_2O_3	CaF_2	1.0×10^{-5}	650	[61]
(V,Ti)N	V_2O_5	5.0×10^{-4}	500	[68]

addition of solid lubricants generally decreases wear rates compared to unlubricated materials. However, performance depends on factors such as lubricant content, matrix material, temperature, and synergistic effects between multiple lubricants.

11.4 TRIBOLOGICAL BEHAVIOR OF SELF-LUBRICATING COMPOSITES

The addition of soft, low shear strength metals, such as silver, gold, and indium, can provide solid lubrication to metal matrix composites, especially at elevated temperatures [74]. In nickel alloy composites made by powder metallurgy, small amounts of silver nanoparticles from 1% to 5% lowered the coefficient of friction from around 0.7 to 0.4 across a temperature range of 25–750 °C. The wear rate remained relatively constant at approximately 0.8×10^{-2} mm^3/m with silver additions [50]. The soft metal reinforcement forms low shear strength smeared films at temperatures above 600 °C that prevent adhesion and reduce friction. However, soft metals may also increase wear rates through abrasive and adhesive mechanisms if smearing does not occur uniformly across the surface [74, 75].

The addition of graphite flakes to metal matrices, especially aluminum alloys, has been extensively studied as a solid lubricant reinforcement. Graphite forms a lubricating tribofilm on the alloy surface during sliding contact that reduces friction and wear. For composites made by stir casting, increasing graphite content from 5 to 15% reduced the coefficient of friction from around 0.8 down to 0.3 [76, 77]. The wear rate decreased initially from 1.2×10^{-2} mm^3/m to 0.8×10^{-2} mm^3/m as the graphitic tribofilm covered more surface area. However, higher graphite contents can reduce mechanical properties and load-bearing capacity, causing wear rate to increase again after the optimal graphite content is exceeded [78]. There is often an optimal graphite content that balances lubrication and mechanical properties. Strategies such as adding hard ceramic particles like SiC have been used to improve wear resistance at higher graphite contents [79]. Graphene nanoplatelets have been explored as reinforcements in aluminum matrix composites, providing remarkable friction and wear improvements. In aluminum alloy composites produced by friction stir processing, adding more graphene decreased the coefficient of friction from 0.5 to 0.2. The wear rate also declined substantially from 1.5×10^{-3} mm^3/m to 0.5×10^{-3} mm^3/m with increasing graphene content [80]. The excellent lubricating ability of graphene nanosheets is attributed to easy shearing and weak bonding between graphene layers that promotes sliding and formation of a transfer film [81]. Graphene also enhances mechanical properties like hardness at low contents before softening the alloy matrix at higher loadings [81, 82].

hBN is commonly added to metal composites like stainless steel for its excellent thermal stability and solid lubrication, even at high temperatures [83]. For stainless steel composites made by powder metallurgy, increasing hBN from 5% to 15% lowered the coefficient of friction from around 0.9 down to 0.55. The wear rate rapidly decreased from 2.1×10^{-2} mm^3/m to 0.7×10^{-2} mm^3/m up to 10% hBN addition [84]. hBN forms a compact tribofilm containing BN, oxides, and sulfides that reduces adhesion and abrasive wear [85]. However, poor wettability between hBN and the metal matrix can reduce interparticle bonding, degrading mechanical properties at higher contents [86]. Applying pressure or coatings to improve hBN dispersion and using alloys like nickel helps maintain properties [52, 87].

For copper-MoS_2 composites by powder metallurgy, the coefficient of friction reduced to 0.15 at 7–10% MoS_2, where it plateaued with further additions. The wear rate exhibited a similar trend, dropping initially from 1.8×10^{-3} mm^3/m to 0.3×10^{-3} mm^3/m up to 10% MoS_2 [87]. The mechanisms involve MoS_2 tribofilms limiting metal-metal contact through easy shear layering [88, 89]. However, MoS_2 decomposes above 650 °C during processing, which can form brittle phases [90]. CaF_2 is utilized to provide solid lubrication in nickel and cobalt alloy composites, especially at high temperatures up to 1000 °C [91, 92]. In laser-clad cobalt alloy composites, increasing CaF_2 content reduced the coefficient of friction from 0.6 down to 0.3. The wear rate also decreased substantially from 1.2×10^{-2} mm^3/m to 0.5×10^{-2} mm^3/m with more CaF_2 [91]. Although CaF_2 decomposes at above 1400 °C, phases like $CaMoO_4$ and $CaCrO_4$ formed from reactions with the matrix still provide effective lubrication [93]. Several oxides, like TiO_2, MoO_3, and Bi_2O_3, are added to metal composites to provide solid lubrication [94–96]. At elevated temperatures, oxides form a glaze layer of low shear strength crystalline phases like Na_2O-B_2O_3 that lubricate sliding interfaces [95]. TiO_2 reduces friction and wear when added to copper composites, attributed to these lubricious oxides formed during sliding. Up to 4% Bi_2O_3 in aluminum composites causes a glaze-like Bi_2O_3 film that lowers the coefficient of friction to 0.47 and reduces wear [94]. However, effectiveness depends on dispersion, temperature, and environmental conditions.

In summary, the addition of solid lubricants, such as graphite, graphene, MoS_2, hBN, WS_2, and CaF_2, to metal matrix composites generally decreases the coefficient of friction and wear rate, especially at lower contents up to 5–10%. The effectiveness varies based on the processing method, test conditions, and matrix material. Table 11.4 summarizes the variation of the coefficient of friction and wear rate for metal matrix composites with different solid lubricants.

Table 11.4 Coefficient of friction and wear rate for metal matrix composites with different solid lubricants

Solid lubricant	*Matrix material*	*Manufacturing method*	*COF*	*Wear rate*	*References*
Graphite	Aluminum alloy	Stir casting	Decreases with increasing graphite content, decrease to ~0.4	Decreases with increasing graphite content up to an optimum, then increases	[76, 77]
Graphite	Aluminum alloy	Powder metallurgy	Decreases with increasing graphite content, decrease to ~0.2	Decreases with increasing graphite content up to 5–10% graphite, then increases	[97]
Graphene	Aluminum alloy	Friction stir processing	Decreases with increasing graphene content, decrease to ~0.25	Decreases with increasing graphene content	[80]
CNTs	Aluminum alloy	Friction stir processing	Decreases with increasing CNT content, decrease to ~0.4	Decreases with increasing CNT content	[80]
MoS_2	Copper	Powder metallurgy	Decreases with increasing MoS_2 content, levels of around 7–10% MoS2 at ~0.15	Decreases with increasing MoS_2 content, then stays relatively constant	[87]

(*Continued*)

Table 11.4 (Continued)

Solid lubricant	*Matrix material*	*Manufacturing method*	*COF*	*Wear rate*	*References*
MoS_2	Stainless steel	Powder metallurgy	Decreases with increasing MoS_2 content, levels of around 10% MoS_2 at ~0.5	Decreases rapidly initially, then stays relatively constant	[98]
hBN	Stainless steel	Powder metallurgy	Decreases with increasing hBN content, levels of around 10% hBN at ~0.6	Decreases rapidly initially, then stays relatively constant	[84]
WS_2	Nickel alloy	Laser cladding	Decreases with increasing WS_2 content	Decreases with increasing WS_2 content, a minimum of around 5–10% WS_2	[92]
CaF_2	Cobalt alloy	Laser cladding	Decreases with increasing CaF_2 content	Decreases with increasing CaF_2 content	[91]

11.5 SELF-LUBRICATING COATINGS

Given the high cost of developing bulk materials with specific properties and the fact that near-surface properties have the greatest influence on tribological behavior, designing thin-film coatings appears to be a prudent choice. This approach not only lowers effective costs but also increases the robustness of mechanical systems by reducing weight, making it a viable strategy for performance optimization. Typically, a thin layer (<10 μm) of solid lubricant is applied strategically to a tribo-pair to lower friction and wear. These solid lubricating thin-film coatings are ideal for applications with a limited number of operating cycles, moderate contact pressure, and low-speed conditions, especially in boundary contact situations [99]. A solid lubricant should be carefully selected to meet the specific operating conditions of the application. The deposition process has a significant impact on the solid lubricant's efficacy and tribological properties. Traditionally, solid lubricating thin films have been produced through methods such as burnishing, spraying, or physical vapor deposition (PVD). In modern applications, magnetron sputtering PVD processes are preferred for developing high-performance coatings. This preference stems from the ability of the magnetron sputtering process to produce high-quality coatings with significantly less contamination since the process takes place inside a vacuum chamber, ensuring a clean environment. Moreover, magnetron-sputtered coatings exhibit higher density, better adhesion, and provide greater flexibility in controlling chemical composition and microstructure, accomplished through co-doping and the addition of reactive gases to the deposition process [100].

One of the most thoroughly researched solid lubricants is MoS_2, which is used in thin films to reduce wear and friction between sliding parts because it exhibits ultra-low friction (up to 0.001) and significantly lower wear in vacuum [101, 102]. Magnetron-sputtered MoS_2 coatings have been successfully used in space mechanisms by NASA and ESA for several decades [14]. Therefore, this section will discuss the use of MoS_2 coatings to achieve ultra-low friction. The low friction coefficient of MoS_2 is attributed to its layered structure, which includes a molybdenum (Mo) layer sandwiched between two sulfur (S) layers connected by covalent bonds. The strong in-plane covalent bonds within MoS_2 are complemented by weak out-of-plane van der Waals interactions, causing mechanical anisotropy. The unique interplay of strong covalent bonds and van der Waals bonding optimizes compressive strength for increased load-bearing capacity while simultaneously reducing shear strength, allowing for easy shearing of the planes for reduced friction [103]. The literature extensively details the low-friction mechanism of MoS_2, offering a comprehensive understanding. At its core, the process hinges on the development of a transfer film on the counter body during sliding and prompts a reorientation within the MoS_2 crystal to a basally oriented structure in both the coating and the transfer film [104]. Figure 11.1 shows the shear-induced reorientation of the MoS_2 crystals during sliding in a dry N_2 test environment.

Sputter-deposited MoS_2 coatings have two distinct morphologies: type I and type II. In type I, the (002) basal planes are perpendicular to the substrate, whereas in type II, they are parallel [105]. Typically, these coatings display a combination of type I and type II morphologies, whereas amorphous structures are also common [106, 107]. During the initial growth phase (about 50 nm), the MoS_2 film is (002) oriented, but subsequent layers become edge-plane oriented [107]. The (002) orientation on the basal planes improves resistance to environmental oxidation, resulting in better friction and wear behavior. This is due to the absence of dangling bonds and reaction sites on the basal planes. In contrast, edge-oriented MoS_2 planes with dangling bonds serve as reactive sites for environmental oxygen. These planes are characterized by a loose columnar structure with higher porosity and length comparable to the thickness of the coating [108]. Exposure to oxygen and water vapor in the air causes lattice distortions and defects, increasing the resistance to sliding. The shorter length of the Mo–O bond, compared to the Mo–S bond, causes atomic-level defects, resulting in the creation of an energy barrier [109]. Higher friction and wear are expected during the initial stages of sliding in these films [110, 111]. However, a reorientation phenomenon occurs during sliding, favoring the alignment of the

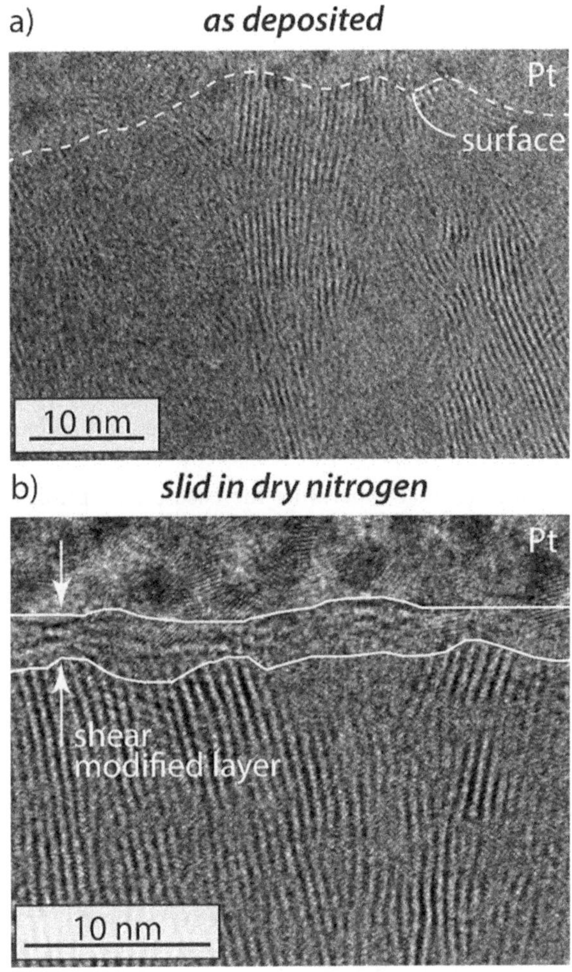

Figure 11.1 TEM images of the coating in its initial deposited state and the surface after sliding under dry N_2 conditions.

(002)-oriented film on the sliding surface, thereby reducing the friction coefficient. Furthermore, sputter-deposited MoS_2 coatings may be sulfur-deficient due to the re-sputtering of small-sized S atoms [112, 113]. Sulfur defects on inert basal planes can also serve as binding sites for oxygen, resulting in the oxidation of MoS_2 [114]. In Figure 11.2(a), the diffraction peaks of MoS_2 are depicted, highlighting the presence of the (002) basal plane and the edge planes (100) and (110). In addition, Figure 11.2(b) illustrates the Raman peaks of MoS_2, specifically capturing the vibrational modes E_{2g} and A_{1g}.

The degradation of MoS_2 is highly dependent on the operating conditions, particularly temperature and humidity levels. At lower temperatures, the primary degradation mechanism in MoS_2 involves the physical adsorption

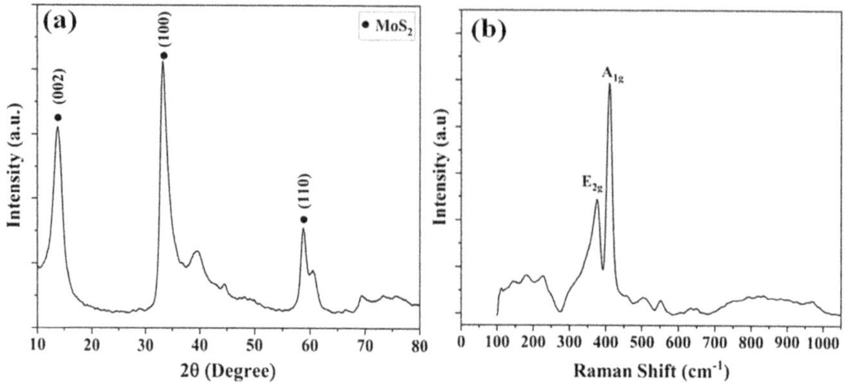

Figure 11.2 (a) GIXRD peaks (b) Raman peaks of as deposited MoS2 film.

of water molecules. As the temperature increases, desorption of water molecules takes place, and a decrease in the friction coefficient is observed. This trend continues until a transition temperature is reached. Upon surpassing this transition temperature, a distinct mechanism takes precedence as thermally activated oxidation becomes dominant [110]. This higher-temperature oxidation process culminates in the conversion of MoS_2 to MoO_3, as exemplified by the following reactions:

$$2MoS_2 + O_2 + 4H_2O = 2MoO_3 + 4H_2S$$

$$2MoS_2 + 9O_2 + 4H_2O = 2MoO_3 + 4H_2SO_4$$

The formation of MoO_3 in these reactions contributes to high abrasiveness below 500 °C, which causes excessive friction, wear, and deterioration of the MoS_2 coating [115]. With increasing temperature, the oxidation rate increases proportionally, resulting in the rapid degradation of MoS_2 [116]. These coatings have limited application in scenarios where temperatures are expected to exceed 350 °C.

Oxidation of MoS_2 is a well-known issue, particularly in humid environments, limiting its wide range of applications. To overcome the inherent ambient air sensitivity of MoS_2 coatings, researchers have strategically incorporated dopants, particularly metals such as Ti, Cr, W, Zr, Ta, Pb, Au, and others [117–126]. These dopants increase coating density (reduce porosity), prevent oxygen penetration, and improve mechanical properties. Metals (oxygen getters) sometimes form sacrificial bonds with oxygen molecules, protecting MoS_2 from oxidation. In Figure 11.3(b), when Zr is doped in the coating, it prefers to react with oxygen instead of MoS_2. The Mo 3d spectrum contains peaks at 229.0 and 232.2 eV, indicating Mo^{4+} in MoS_2, and at 232.7 and 235.8 eV, corresponding to MoO_3. Interestingly, after etching, the MoO_3 peaks vanish, implying that Zr plays an important role in preventing MoS_2 oxidation. The Zr 3d spectrum in Figure 11.3(c) shows

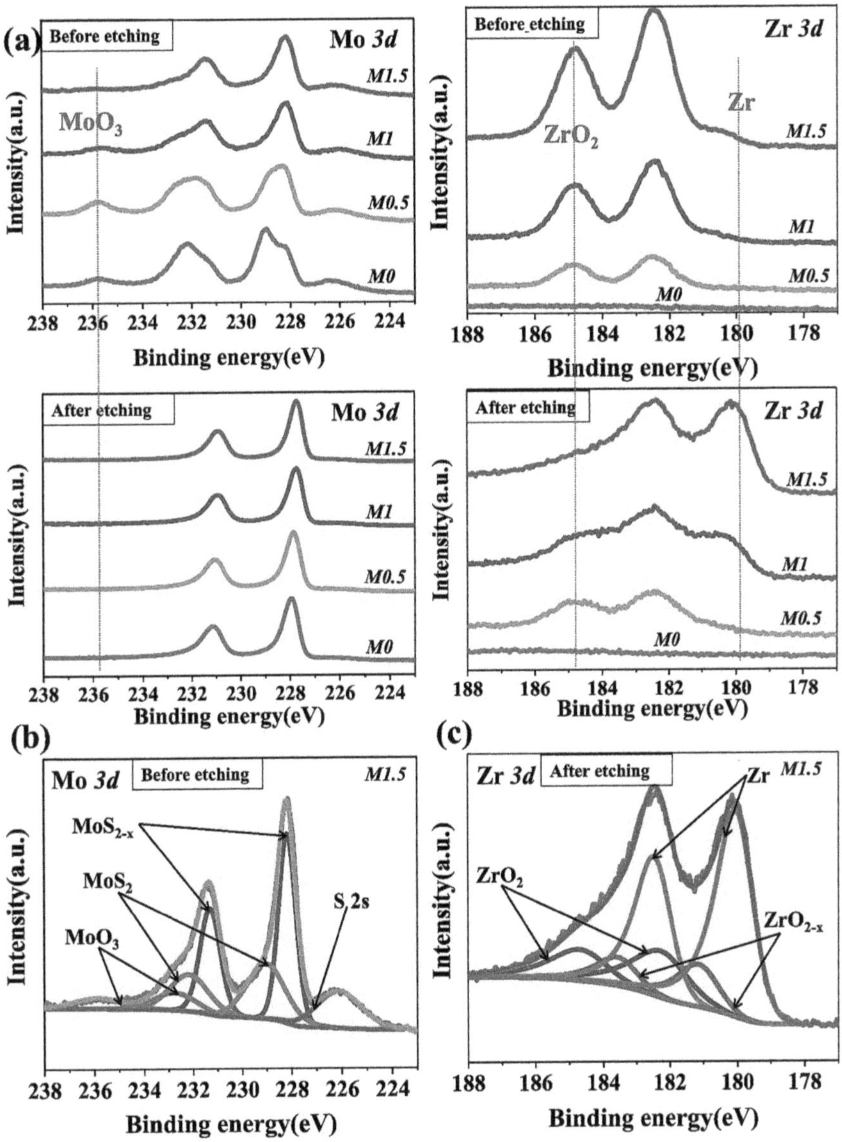

Figure 11.3 (a) XPS spectra show Mo 3d and Zr 3d for MoS_2 (M0) and MoS_2/Zr coatings (M0.5, M1, M1.5) before and after Ar+ etching. (b) Mo 3d breakdown for MoS_2/Zr coating (M1.5) is shown pre-etching, and (c) Zr 3d breakdown post-etching. Number after M shows the current of Zr target in Ampere used for achieving different Zr contents.

that the peaks at 182.3 and 184.7 eV correspond to ZrO_2, indicating the affinity of Zr for reacting with oxygen. Selective reactivity of Zr inhibits the oxidation of MoS_2 within the coating, making it an important factor in maintaining the lubricating characteristics of the coating material [117]. Beyond metals, nonmetals such as C and N have been investigated for their ability to improve coating compactness and hardness, frequently resulting in amorphous structures or a matrix embedding MoS_2 nanocrystals [127–132]. Figure 11.4 shows the SEM images of Mo-S-N coatings doped with different concentrations of N by changing the N_2 flow in the deposition chamber. The pristine MoS_2 (Figure 11.4(a, d)) exhibits high porosity and features dendrite columns along the thickness of the coating. This structure may lead to the diffusion of oxygen and water vapor into the coating more easily. However, as the concentration of N increases, the density improves significantly, and the dendritic column disappears, and the morphology becomes featureless [133]. The introduction of dopants has been observed to exhibit a slight elevation in the friction coefficient when compared to pristine MoS_2. However, this unfavorable effect is counterbalanced by a remarkable enhancement in wear life, particularly in humid air and elevated temperatures. Intriguingly, this positive outcome is not consistently replicated in vacuum conditions, where the incorporation of dopants appears to diminish the wear life. Figure 11.5 depicts the friction curves for MoS_2 films doped with varying Cr concentrations, obtained by adjusting the target current from 0 to 0.4 A. These doped films exhibit contrasting performance in vacuum (Figure 11.5(a)) and atmospheric conditions (Figure 11.5(b)). Notably,

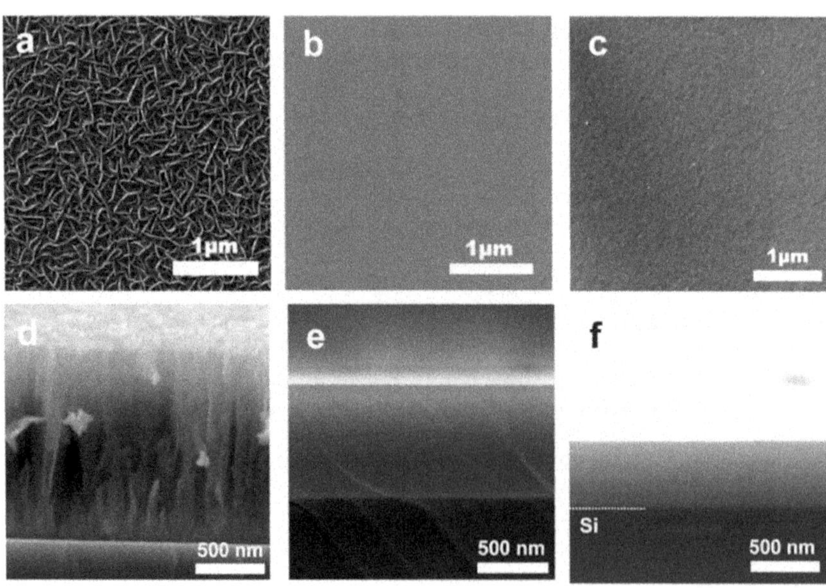

Figure 11.4 SEM topographical and cross-sectional images of MoS_2 films modified by N_2 reactive sputtering (a and d) 0 sccm (b and e) 5 sccm (c and f) 20 sccm of N_2.

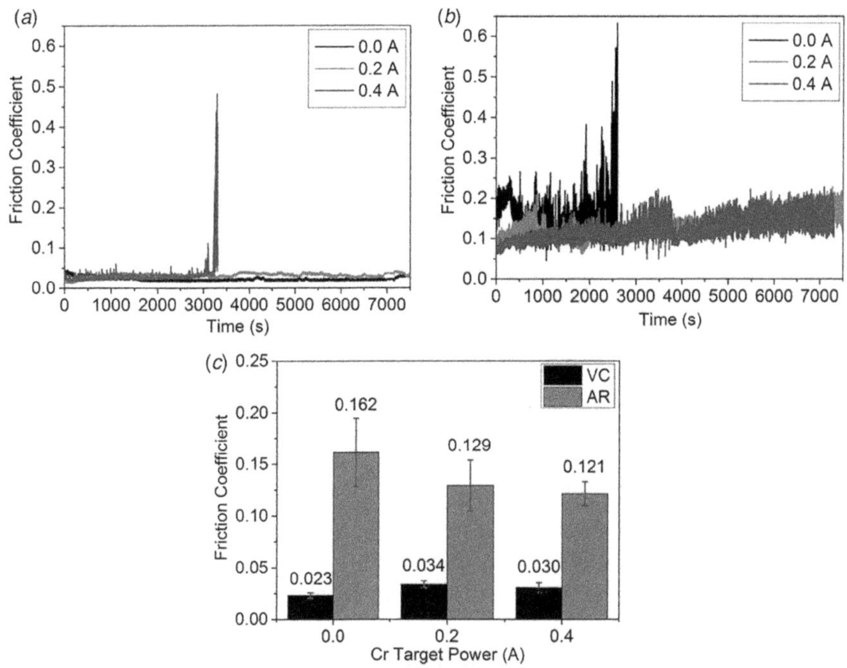

Figure 11.5 The variation in friction coefficient over time is depicted for MoS_2 films doped with Cr at target current of 0, 0.2 A, and 0.4 A, subjected to (a) vacuum (VC), (b) ambient air (AR) environments and(c) mean friction coefficient.

Cr doping is associated with increased friction, resulting in sudden failure at higher Cr concentrations in vacuum environments. In contrast, under atmospheric conditions, the same doping process significantly increases wear life while decreasing the friction coefficient [134].

While doping has traditionally been a popular method for increasing the oxidation resistance of MoS_2, other approaches have emerged. These include the development of multilayer films, the creation of fullerene-like nanostructures, and the application of chameleon coatings, representing diverse strategies beyond conventional doping for improving MoS_2 oxidation resistance. The development of multilayer interfaces may impede the growth of the porous structure in MoS_2 films. This structure frequently promotes a preferred (002) orientation parallel to the substrate, resulting in lower coefficients of friction and increased humidity resistance. Furthermore, these interfaces act as barriers to dislocation motion, creating an energetically unfavorable environment for dislocation generation as the bilayer period and crystallite size approach nanometer dimensions. As a result, the multilayer film maintains its desirable mechanical and tribological properties [135–137]. Chhowalla and Amaratunga showed that thin films of hollow MoS_2 nanoparticles, deposited through a localized high-pressure arc discharge method, display significantly lower friction levels, outperforming sputtered MoS_2 films by an order of magnitude. The unique dry behavior in humid conditions is attributed to the presence of curved S–Mo–S planes, effectively preventing oxidation [138].

Chameleon coatings use nanoengineering techniques to develop nanocomposite coatings that can adjust to changing environmental conditions. Through the integration of multiple solid lubricants, these coatings autonomously release suitable lubricant materials and direct nanoscale volumes to the contact zone, resulting in macroscopic solid lubricating films as required. These coatings, inspired by the adaptive features of chameleon lizards, self-regulate surface chemistry and microstructure, ensuring resilience to changing conditions while maintaining consistent and effective tribological performance across a wide range of environments and temperatures [108]. The YSZ/Au/MoS_2/DLC chameleon coating exhibits an adaptive environmental behavior, with hard YSZ phases in the Au matrix enveloping MoS_2 and DLC reservoirs (Figure 11.6(a)). This multifunctional coating structure and chemistry enable effective lubrication in diverse conditions: DLC and MoS_2 provide lubrication in humid air and inert atmospheres, respectively, while Au contributes to lubrication at high temperatures [139] (Figure 11.6(b)). Similarly, numerous chameleon coatings, such as Al_2O_3/Au/

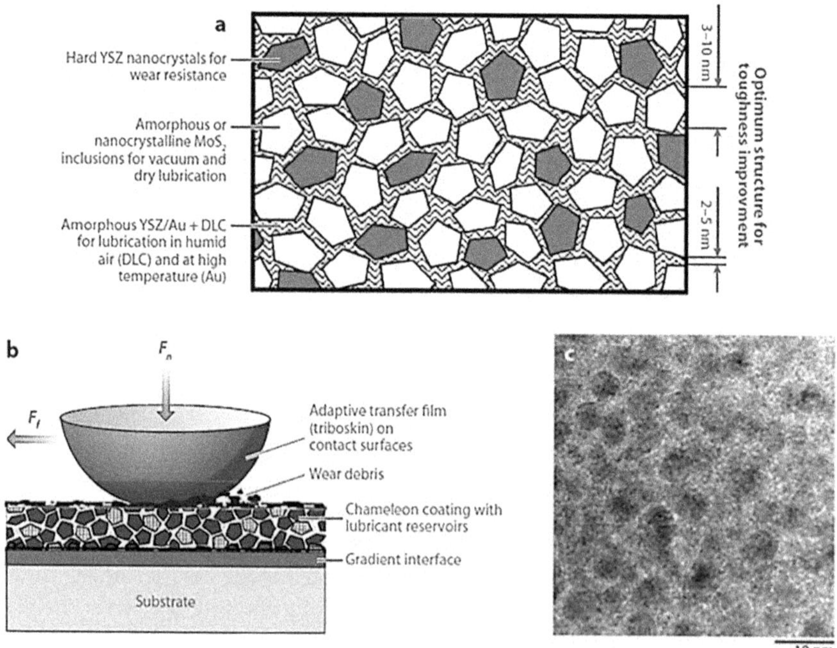

Figure 11.6 Schematic of a YSZ-MoS_2-DLC-Au coating: (a) microstructural design (b) sliding response under various ambient conditions. (c) Microscopic image of the coating reveals MoS_2 inclusions within a nanocrystalline-amorphous YSZ-Au matrix.

MoS_2/DLC, graphite–MoS_2, and H–DLC/DLC/MoS_2, have been effectively engineered to adjust to various operational environments [116, 140, 141].

Therefore, thin-film coatings, especially those incorporating solid lubricants like MoS_2 deposited through magnetron sputtering, offer a cost-effective means to enhance tribological performance. Despite the success of MoS_2, challenges like oxidation have led to innovative solutions such as doping with metals and nonmetals. Alternative approaches, including multilayer films and chameleon coatings inspired by adaptive features in nature, aim to further improve oxidation resistance and adaptability to diverse environmental conditions. Careful consideration of operating conditions and continued research into materials and deposition techniques will be crucial for advancing the effectiveness and versatility of tribological coatings in various mechanical systems.

11.6 INDUSTRIAL APPLICATIONS OF SOLID LUBRICANTS

Solid lubricants are being increasingly utilized in self-lubricating composites and coatings across diverse industrial sectors to provide adaptive lubrication under extreme conditions where liquid lubricants fail [142]. Soft metals, including silver, when added at 5–15 wt% to alloys like TiAl used extensively in aerospace bearings and seals, offer lubrication up to 500 °C by forming adaptive tribofilms [6]. Transition metal dichalcogenides, predominantly MoS_2, are applied as low-friction coatings on aerospace components owing to their ability to provide friction coefficients as low as 0.006 in vacuum environments [143]. MoS_2 enables high-temperature lubrication exceeding 350 °C in automotive engines and metal-forming applications under pressures above 1 GPa due to its lamellar structure [14].

Oxides are also effective solid lubricants, with combinations of TiO_2–Bi_2O_3 and binary oxides like V_2O_5 providing low friction from 500 to 1000 °C in hot forming dies, cutting tools, and wire drawing dies because of their high-temperature stability [144]. Fluorides, including BaF_2 and CaF_2, facilitate self-lubrication above 500 °C in engine components and exhibit friction coefficients below 0.25 at 1000 °C in cutting tools and hot bearings as a result of their ductility at high temperatures [145]. Polymers like polytetrafluoroethylene (PTFE) demonstrate low-friction properties with coefficients below 0.1 at cryogenic conditions in space mechanisms and dry gas seals due to their high hardness [146]. Graphite is widely adopted in polymer composites for bushings and bearings, where it provides adaptive lubrication in humid environments, yielding a 50% decrease in friction coefficient compared to unfilled polymers because of its laminar structure [147].

Dry-film lubricants are also used to protect metal surfaces from damage during transportation, handling, and storage. They can be used to provide lubrication during metalworking operations. Dry-film lubricants are typically petroleum-based, but bio-based lubricants are also available. Bio-based lubricants offer several advantages over petroleum-based lubricants, including a cleaner work environment, lower worker health risks, and significant cost savings. Lubricants are available in liquid, solid, and gaseous forms. SLs are more effective than liquid lubricants in high-load or high-speed applications. They are also used in applications where conventional lubricants cannot be used, such as extreme pressure or temperature conditions. In the automotive industry, sheet metal forming depends on the utilization of both liquid and dry-film lubricants to achieve optimal performance. Liquid lubricants find common usage in the press shop, while dry-film lubricants are applied in the rolling mill. Specifically, water-free dry-film lubricants, also known as hot-melts, are administered onto sheet metal at rolling mills. This application serves to enhance drawing performance and provide corrosion protection for the construction of automotive bodies. Dry lubricants offer several advantages over conventional oil lubricants [148], including reduced lubricant consumption, a wider process window, elimination of die washing, consistent coating thickness, exceptional deep-drawing performance, corrosion resistance, compatibility with assembly operations and painting, and minimal environmental impact. Solid lubricants, such as E1 from Zeller and Gmelin, and environmentally friendly alternatives like wheat flour [149] and starch–oil composites, also play a crucial role in metalworking, particularly in non-cutting forming, cold and hot forging, pressure die-casting and drawing, and cold extrusion operations, providing effective separation between tool and workpiece and protecting sheet metal integrity.

Solid lubricants enable self-lubricating composites and coatings to perform under demanding conditions across key industrial sectors, including aerospace, automotive, metal forming, and energy. Their unique structures and properties allow adaptive lubrication from cryogenic to high temperatures.

11.7 FUTURE SCOPE AND OPPORTUNITIES

There are several promising approaches to improve solid lubricants, expand their applications, and develop self-lubricating materials for extreme conditions. Using combinations of multiple solid lubricants can extend the operating temperature range compared to individual additives. For example, silver and fluorides together enable lubrication from 25 to 1000 °C in Ni-based composites.

Another scope for innovation is developing multifunctional self-lubricating composites with enhanced mechanical, thermal, and electrical properties for demanding applications through careful selection of matrix, reinforcement, and fabrication method. For instance, self-lubricating ceramics with high hardness and fracture toughness are attractive for cutting tools and bearings. Integrating solid lubricants into lightweight metal matrices can give composites suitable for automotive and aerospace components. Opportunities exist for designing interfaces between the matrix and solid lubricants in self-lubricating composites, through coatings and physicochemical treatments for stronger interfacial bonding and load transfer. This can potentially improve wear resistance. Exploring ways to replenish lubricants during service for extended component life is another area worth investigating. Computational modeling approaches can accelerate composite formulation and optimization as well as predict longevity more efficiently.

In summary, exploring novel solid lubricants, devising multifunctional smart composites, engineering interfaces and microstructures, enabling lubricant replenishment, and utilizing advanced tools for characterization and modeling offer considerable scope to develop the next generation of self-lubricating materials tailored for extreme environment applications.

11.8 CONCLUSION

In conclusion, solid lubricants offer unique capabilities to address the lubrication challenges of the 21st century. With continued material innovation and manufacturing improvements, solid lubricants have excellent prospects for enabling more sustainable, reliable, and durable operation of mechanical systems working in harsh conditions. Their adaptive lubrication properties make them highly promising for mitigating friction and wear issues in extreme environments. By extending equipment lifetime and efficiency, solid lubricant materials will be vital in reducing energy consumption across key industries such as transportation, manufacturing, and power generation.

REFERENCES

[1] J.-H. Ouyang, C.-C. Shi, Z.-G. Liu, Y.-M. Wang, and Y.-J. Wang, *Wear* **330**, 272 (2015).

[2] C. Donnet and A. Erdemir, *Tribol Lett* **17**, 389 (2004).

[3] E.R. Booser, *Tribology Data Handbook: An Excellent Friction, Lubrication, and Wear Resource*. CRC Press **156**, (1997).

[4] J. Li, D. Xiong, Z. Huang, J. Kong, and J. Dai, *Wear* **267**, 576 (2009).

[5] S.S. Akhtar, *Ceram Int* **47**, 20745 (2021).

[6] X. Shi, Z. Xu, M. Wang, W. Zhai, J. Yao, S. Song, A. Qamar ud Din, and Q. Zhang, *Wear* **303**, 486 (2013).

[7] W. Duan, Y. Sun, C. Liu, S. Liu, Y. Li, C. Ding, G. Ran, and L. Yu, *Tribol Int* **95**, 324 (2016).

[8] J. Zhen, S. Zhu, J. Cheng, M. Li, Y. Lu, Z. Qiao, and J. Yang, *Tribol Int* **114**, 322 (2017).

[9] Y. Wang, Y. Gao, Y. Li, M. Li, L. Sun, W. Zhai, and K. Li, *Wear* **444–445**, 203140 (2020).
[10] M.H. Cho, J. Ju, S.J. Kim, and H. Jang, *Wear* **260**, 855 (2006).
[11] W. Huai, C. Zhang, and S. Wen, *Friction* **9**, 1660 (2021).
[12] X. Shi, W. Zhai, M. Wang, Z. Xu, J. Yao, S. Song, and Y. Wang, *Wear* **310**, 1 (2014).
[13] L. Kong, Q. Bi, M. Niu, S. Zhu, J. Yang, and W. Liu, *Tribol Int* **64**, 53 (2013).
[14] M.R. Vazirisereshk, A. Martini, D.A. Strubbe, and M.Z. Baykara, *Lubricants* 7, (2019).
[15] S. Sukumaran, L. Francis Xavier, and B. Deepanraj, *Mater Today Proc* 77, 478 (2023).
[16] C. Shekhar, M.F. Wani, and R. Sehgal, *Wear* **526–527**, 204890 (2023).
[17] C. Li, L. Duan, S. Tan, W. Zhang, and B. Pan, *Int J Refract Met Hard Mater* 75, 118 (2018).
[18] C.K. Sahoo and M. Masanta, *J Mater Process Technol* **243**, 229 (2017).
[19] L. Kong, Q. Bi, M. Niu, S. Zhu, J. Yang, and W. Liu, *J Eur Ceram Soc* **33**, 51 (2013).
[20] T. Murakami, J.H. Ouyang, S. Sasaki, K. Umeda, and Y. Yoneyama, *Wear* **259**, 626 (2005).
[21] J.H. Ouyang, S. Sasaki, T. Murakami, and K. Umeda, *Mater Sci Eng A* **386**, 234 (2004).
[22] C. Shekhar, M.F. Wani, and R. Sehgal, *Tribol Int* **188**, 108884 (2023).
[23] M. Valefi, M. de Rooij, D.J. Schipper, and L. Winnubst, *J Eur Ceram Soc* **32**, 2235 (2012).
[24] S. Zhu, Q. Bi, M. Niu, J. Yang, and W. Liu, *Wear* **274–275**, 423 (2012).
[25] S. Zhu, J. Cheng, Z. Qiao, Y. Tian, and J. Yang, *J Tribol* **138**, (2016).
[26] L.-M. Berger, C.C. Stahr, S. Saaro, S. Thiele, M. Woydt, and N. Kelling, *Wear* **267**, 954 (2009).
[27] N. Fateh, G.A. Fontalvo, G. Gassner, and C. Mitterer, *Wear* **262**, 1152 (2007).
[28] E. Liu, W. Wang, Y. Gao, and J. Jia, *Tribol Lett* **47**, 21 (2012).
[29] F. Ye, Z. Lou, Y. Wang, and W. Liu, *Tribol Int* **167**, 107402 (2022).
[30] H. Torres, S. Slawik, C. Gachot, B. Prakash, and M. Rodríguez Ripoll, *Surf Coat Technol* **337**, 24 (2018).
[31] R. Kumar, M. Antonov, Y. Holovenko, and A. Surzenkov, *Tribol Lett* **68**, 51 (2020).
[32] J. Chen, Y. An, J. Yang, X. Zhao, F. Yan, H. Zhou, and J. Chen, *Surf Coat Technol* **235**, 521 (2013).
[33] A.A. Voevodin, C. Muratore, and S.M. Aouadi, *Surf Coat Technol* **257**, 247 (2014).
[34] S.M. Aouadi, Y. Paudel, B. Luster, S. Stadler, P. Kohli, C. Muratore, C. Hager, and A.A. Voevodin, *Tribol Lett* **29**, 95 (2008).
[35] Y. Zhou, Y. Dong, H. Yin, Z. Li, R. Yan, D. Li, Z. Gu, X. Sun, L. Shi, and Z. Zhang, *Sci Rep* **8**, 13400 (2018).
[36] T. Kim, D. Singh, and M. Singh, *Energy Procedia* **69**, 900 (2015).
[37] H. Tan, S. Wang, Y. Yu, J. Cheng, S. Zhu, Z. Qiao, and J. Yang, *Tribol Int* **122**, 228 (2018).
[38] J. Lu, S. Yang, J. Wang, and Q. Xue, *Wear* **249**, 1070 (2001).
[39] D. Berman, A. Erdemir, and A. V. Sumant, *Mater Today* **17**, 31 (2014).

[40] P.D. Srivyas, M.F. Wani, R. Sehgal, C.S. Singh Bisht, M.S. Charoo, A. Raina, and M. Irfan Ul Haq, *Tribol Int* **168**, 107434 (2022).
[41] F.A. Essa, A.H. Elsheikh, J. Yu, O.A. Elkady, and B. Saleh, *J Mater Res Technol* **12**, 283 (2021).
[42] Z. Xu, Q. Zhang, P. Jing, and W. Zhai, *Tribol Lett* **58**, 3 (2015).
[43] W.Z.K.Y. Yecheng, X.X. Shi and J. Yao, *Tribol Trans* **58**, 729 (2015).
[44] M.X. Wei, S.Q. Wang, and X.H. Cui, *Wear* **274–275**, 84 (2012).
[45] S. Zhang, J. Zhou, B. Guo, H. Zhou, Y. Pu, and J. Chen, *Mater Sci Eng A* **491**, 47 (2008).
[46] B. Podgornik, T. Kosec, A. Kocijan, and Č. Donik, *Tribol Int* **81**, 267 (2015).
[47] X.-L. Lu, X.-B. Liu, P.-C. Yu, S.-J. Qiao, Y.-J. Zhai, M.-D. Wang, Y. Chen, and D. Xu, *Opt Laser Technol* **78**, 87 (2016).
[48] H. Torres, B. Podgornik, M. Jovičević-Klug, and M. Rodríguez Ripoll, *Wear* **490–491**, 204187 (2022).
[49] R. Kumar, S. Aydinyan, R. Ivanov, L. Liu, M. Antonov, and I. Hussainova, *Materials* **15**, (2022).
[50] R. Tyagi, D.S. Xiong, J. Li, and J. Dai, *Wear* **269**, 884 (2010).
[51] X. Zhu, X. Wei, Y. Huang, F. Wang, and P. Yan, *Metals (Basel)* **9**, (2019).
[52] L. Du, C. Huang, W. Zhang, T. Li, and W. Liu, *Surf Coat Technol* **205**, 3722 (2011).
[53] K.P. Furlan, J.D.B. de Mello, and A.N. Klein, *Tribol Int* **120**, 280 (2018).
[54] S.C.E.L.H. Du Weiqi Jing Shuangming Du and H. Cai, *Tribol Trans* **65**, 66 (2022).
[55] V.M. Gopinath and S. Arulvel, *Mater Today Proc* **43**, 817 (2021).
[56] M. Rodríguez Ripoll, A.M. Tomala, L. Pirker, and M. Remškar, *Tribol Lett* **68**, 41 (2020).
[57] P. Afanasiev and C. Lorentz, *J Phys Chem C* **123**, 7486 (2019).
[58] S.-H. Kim and S. Wohn Lee, *Ceram Int* **40**, 779 (2014).
[59] R. Kumar, M. Antonov, U. Beste, and D. Goljandin, *Int J Refract Met Hard Mater* **86**, 105126 (2020).
[60] L. Kong, Q. Bi, S. Zhu, J. Yang, and W. Liu, *Tribol Int* **45**, 43 (2012).
[61] Y. Jin, K. Kato, and N. Umehara, *Tribol Lett* **4**, 243 (1998).
[62] S. Mazumder, H.S.C. Metselaar, N.L. Sukiman, and N.W.M. Zulkifli, *J Eur Ceram Soc* **40**, 4974 (2020).
[63] J.H. Ouyang, Y.F. Li, Y.M. Wang, Y. Zhou, T. Murakami, and S. Sasaki, *Wear* **267**, 1353 (2009).
[64] M. Erkin Cura, S.-H. Kim, T. Muukkonen, S. Varjus, A. Vaajoki, O. Söderberg, T. Suhonen, U. Kanerva, S. Wohn Lee, and S.-P. Hannula, *Ceram Int* **39**, 2093 (2013).
[65] R. Kumar, B. Malaval, M. Antonov, and G. Zhao, *Tribol Int* **147**, 106282 (2020).
[66] R. Kumar, L. Liu, M. Antonov, R. Ivanov, and I. Hussainova, *Materials* **14**, (2021).
[67] G. Gassner, P.H. Mayrhofer, K. Kutschej, C. Mitterer, and M. Kathrein, *Surf Coat Technol* **201**, 3335 (2006).
[68] J.H. Ouyang, T. Murakami, and S. Sasaki, *Wear* **263**, 1347 (2007).
[69] M.E. Cura, X.W. Liu, U. Kanerva, S. Varjus, A. Kivioja, O. Söderberg, and S.-P. Hannula, *Tribol Int* **87**, 23 (2015).

[70] M.E. Cura, M. Trebala, Y. Ge, P. Klimczyk, and S.-P. Hannula, *Wear* **478–479**, 203920 (2021).
[71] M. Niu, Q. Bi, S. Zhu, J. Yang, and W. Liu, *J Alloys Compd* **555**, 367 (2013).
[72] X. Shi, S. Song, W. Zhai, M. Wang, Z. Xu, J. Yao, A. Qamar ud Din, and Q. Zhang, *Mater Des* **55**, 75 (2014).
[73] Z. Xu, Q. Zhang, and W. Zhai, *RSC Adv* **5**, 45044 (2015).
[74] J.R. Lince, *Tribol Lett* **17**, 419 (2004).
[75] P. Stoyanov, H.W. Strauss, and R.R. Chromik, *Wear* **274–275**, 149 (2012).
[76] F. Akhlaghi and S.A. Pelaseyyed, *Mater Sci Eng A* **385**, 258 (2004).
[77] F. Akhlaghi and A. Zare-Bidaki, *Wear* **266**, 37 (2009).
[78] F. Akhlaghi and S. Mahdavi, in *Advances in Materials and Processing Technologies II* (Trans Tech Publications Ltd, 2011), pp. 1878–1886.
[79] S. Suresha and B.K. Sridhara, *Compos Sci Technol* **70**, 1652 (2010).
[80] R. Maurya, B. Kumar, S. Ariharan, J. Ramkumar, and K. Balani, *Mater Des* **98**, 155 (2016).
[81] B. Xue, Q. Zhu, X. Shi, W. Zhai, K. Yang, and Y. Huang, *J Mater Eng Perform* **25**, 4126 (2016).
[82] H. Algul, M. Tokur, S. Ozcan, M. Uysal, T. Cetinkaya, H. Akbulut, and A. Alp, *Appl Surf Sci* **359**, 340 (2015).
[83] S. Mahathanabodee, T. Palathai, S. Raadnui, R. Tongsri, and N. Sombatsompop, in *Multi-Functional Materials and Structures IV* (Trans Tech Publications Ltd, 2013), pp. 307–310.
[84] S. Mahathanabodee, T. Palathai, S. Raadnui, R. Tongsri, and N. Sombatsompop, *Wear* **316**, 37 (2014).
[85] S. Mahathanabodee, T. Palathai, S. Raadnui, R. Tongsri, and N. Sombatsompop, *Mater Des* (1980-2015) **46**, 588 (2013).
[86] S. Mahathanabodee, T. Palathai, S. Raadnui, R. Tongsri, and N. Sombatsompop, in *Processing and Fabrication of Advanced Materials* (Trans Tech Publications Ltd, 2012), pp. 216–219.
[87] L. Du, W. Zhang, W. Liu, and J. Zhang, *Surf Coat Technol* **205**, 2419 (2010).
[88] Y. Zhang, J. Michael Shockley, P. Vo, and R.R. Chromik, *Tribol Lett* **62**, 9 (2016).
[89] A.M. Kovalchenko, O.I. Fushchich, and S. Danyluk, *Wear* **290–291**, 106 (2012).
[90] H. Kato, M. Takama, Y. Iwai, K. Washida, and Y. Sasaki, *Wear* **255**, 573 (2003).
[91] H. Yan, J. Zhang, P. Zhang, Z. Yu, C. Li, P. Xu, and Y. Lu, *Surf Coat Technol* **232**, 362 (2013).
[92] M.-S. Yang, X.-B. Liu, J.-W. Fan, X.-M. He, S.-H. Shi, G.-Y. Fu, M.-D. Wang, and S.-F. Chen, *Appl Surf Sci* **258**, 3757 (2012).
[93] S. Zhu, Q. Bi, H. Wu, J. Yang, and W. Liu, *Tribol Lett* **41**, 535 (2011).
[94] J.-O. Kliemann, H. Gutzmann, F. Gärtner, H. Hübner, C. Borchers, and T. Klassen, *J Therm Spray Technol* **20**, 292 (2011).
[95] G.M. Hamilton, *Proc Inst Mech Eng C J Mech Eng Sci* **197**, 53 (1983).
[96] B. Yao, Z. Han, and K. Lu, *Wear* **301**, 608 (2013).
[97] F. Akhlaghi and S. Mahdavi, *Adv Mater Res* **264**, 1878 (2011).
[98] S. Raadnui, S. Mahathanabodee, and R. Tongsri, *Wear* **265**, 546 (2008).
[99] J.R. Lince, *Lubricants* **8**, 74 (2020).

[100] S.H. Mukhtar, A. Gulzar, S. Saleem, M.F. Wani, R. Sehgal, A.A. Yakovenko, I.G. Goryacheva, and M.D. Sharma, *Tribol Int* **192**, 109194 (2024).
[101] Z. Chen, X. He, C. Xiao, and S.H. Kim, *Lubricants* **6** (2018).
[102] C. Donnet, T. Le Mogne, and J.M. Martin, *Surf Coat Technol* **62**, 406 (1993).
[103] O. Samy, S. Zeng, M.D. Birowosuto, and A. El Moutaouakil, *Crystals (Basel)* **11**, (2021).
[104] J.-M. Martin, in *Superlubricity*, edited by A. Erdemir and J.-M. Martin (Elsevier Science B.V., Amsterdam, 2007), pp. 207–225.
[105] P.D. Fleischauer, *A S L E Trans* **27**, 82 (1984).
[106] S.H. Mukhtar, M.F. Wani, R. Sehgal, and M.D. Sharma, *Tribol Int* **178**, 108017 (2023).
[107] T. Spalvins, *Thin Solid Films* **73**, 291 (1980).
[108] C. Muratore and A.A. Voevodin, *Annu Rev Mater Res* **39**, 297 (2009).
[109] J.R. Lince, M.R. Hilton, and A.S. Bommannavar, *Thin Solid Films* **264**, 120 (1995).
[110] H.S. Khare and D.L. Burris, *Tribol Lett* **52**, 485 (2013).
[111] M. Tagawa, J. Ikeda, H. Kinoshita, M. Umeno, and N. Ohmae, in *Protection of Space Materials from the Space Environment*, edited by J.I. Kleiman and R.C. Tennyson (Springer Netherlands, Dordrecht, 2001), pp. 73–84.
[112] X. Zhang, W. Lauwerens, J. He, and J.-P. Celis, *J Vac Sci Technol A* **21**, 416 (2003).
[113] W. Tillmann, A. Wittig, D. Stangier, H. Moldenhauer, C.-A. Thomann, J. Debus, D. Aurich, and A. Bruemmer, *Surf Coat Technol* **385**, 125358 (2020).
[114] J.F. Curry, N. Argibay, T. Babuska, B. Nation, A. Martini, N.C. Strandwitz, M.T. Dugger, and B.A. Krick, *Tribol Lett* **64**, 11 (2016).
[115] M. Buttery, S. Lewis, A. Kent, R. Bingley, and M. Cropper, *Lubricants* **8**, (2020).
[116] C.C. Baker, J.J. Hu, and A.A. Voevodin, *Surf Coat Technol* **201**, 4224 (2006).
[117] L. Li, Z. Lu, J. Pu, and B. Hou, *Appl Surf Sci* **541**, 148453 (2021).
[118] X. Wang, Y. Xing, S. Ma, X. Zhang, K. Xu, and D.G. Teer, *Surf Coat Technol* **201**, 5290 (2007).
[119] Y. Chen, J. Sun, Y. Liu, Q. Li, S. Xiao, and F. Su, *Surf Coat Technol* **452**, 129135 (2023).
[120] H. Li, G. Zhang, and L. Wang, *Wear* **350–351**, 1 (2016).
[121] X. Gao, M. Hu, Y. Fu, L. Weng, W. Liu, and J. Sun, *J Alloys Compd* **815**, 152483 (2020).
[122] T.A. Mufti, S.G. Jan, M.F. Wani, and R. Sehgal, *Tribol Int* **182**, 108374 (2023).
[123] P. Serles, E. Nicholson, J. Tam, N. Barri, J.-B. Chemin, G. Wang, Y. Michel, C.V. Singh, P. Choquet, A. Saulot, T. Filleter, and G. Colas, *Adv Funct Mater* **32**, 2110429 (2022).
[124] C. Zeng, J. Pu, H. Wang, S. Zheng, L. Wang, and Q. Xue, *Ceram Int* **45**, (2019).
[125] W.D. Sun, J. Wang, K.W. Wang, J.J. Pan, R. Wang, M. Wen, and K. Zhang, *Surf Coat Technol* **422**, 127490 (2021).
[126] J. Tian, J. Jin, C. Zhang, J. Xu, W. Qi, Q. Yu, W. Deng, Y. Wang, X. Li, X. Chen, and L. Ma, *Appl Surf Sci* **578**, 152068 (2022).
[127] T. Takeno, S. Abe, K. Adachi, H. Miki, and T. Takagi, *Diam Relat Mater* **19**, 548 (2010).

[128] J. V. Pimentel, T. Polcar, and A. Cavaleiro, *Surf Coat Technol* **205**, 3274 (2011).
[129] A. Seynstahl, M. Köbrich, T. Rosnitschek, M. Göken, and S. Tremmel, *Surf Coat Technol* **477**, 130343 (2024).
[130] K. Hebbar Kannur, T. Huminiuc, T. Bin Yaqub, T. Polcar, C. Pupier, C. Héau, and A. Cavaleiro, *Surf Coat Technol* **408**, 126791 (2021).
[131] K. Hebbar Kannur, T. Bin Yaqub, C. Pupier, C. Héau, and A. Cavaleiro, *ACS Appl Mater Interfaces* **12**, 43299 (2020).
[132] L. Gu, P. Ke, Y. Zou, X. Li, and A. Wang, *Appl Surf Sci* **331**, 66 (2015).
[133] T. Hudec, M. Mikula, L. Satrapinskyy, T. Roch, M. Truchlý, P. Švec, T. Huminiuc, and T. Polcar, *Appl Surf Sci* **486**, 1 (2019).
[134] J. Liu, X. Sui, Z. Yan, G. Huang, and J. Hao, *J Tribol* **143**, (2021).
[135] S. Watanabe, J. Noshiro, and S. Miyake, *Surf Coat Technol* **183**, 347 (2004).
[136] Y. Fu, T. He, W. Yang, J. Xu, B. Mu, X. Pang, and P. Wang, *Coatings* **9**, (2019).
[137] S. Banday and M.F. Wàñi, *Surf Interface Anal* **51**, 649 (2019).
[138] M. Chhowalla and G.A.J. Amaratunga, *Nature* **407**, 164 (2000).
[139] A. Voevodin, T. Fitz, J.J. Hu, and J. Zabinski, *J Vac Sci Technol A* **20**, 1434 (2002).
[140] H. Han, Z. Qian, F. Meng, and Z. Cui, *Proc Instit Mech Eng J: J Eng Tribol* **233**, 1888 (2019).
[141] G. Yu, Z. Gong, B. Jiang, D. Wang, C. Bai, and J. Zhang, *Diam Relat Mater* **102**, 107668 (2020).
[142] S. Zhu, J. Cheng, Z. Qiao, and J. Yang, *Tribol Int* **133**, 206 (2019).
[143] W. Zhai, X. Shi, M. Wang, Z. Xu, J. Yao, S. Song, and Y. Wang, *Wear* **310**, 33 (2014).
[144] S. Zhu, Q. Bi, J. Yang, W. Liu, and Q. Xue, *Tribol Int* **44**, 445 (2011).
[145] J. Zhen, J. Cheng, H. Tan, Q. Sun, S. Zhu, J. Yang, and W. Liu, *Friction* **9**, 990 (2021).
[146] Q. Wang, F. Zheng, and T. Wang, *Cryogenics (Guildf)* **75**, 19 (2016).
[147] L. Chang, Z. Zhang, H. Zhang, and K. Friedrich, *Tribol Int* **38**, 966 (2005).
[148] M. Meiler, M. Pfestorf, M. Geiger, and M. Merklein, *Wear* **255**, 1455 (2003).
[149] H. Yoshimura, S. Torikai, T. Nishihara, T. Nonishi, and N. Inouchi, *J Mater Process Technol* **125–126**, 375 (2002).

Chapter 12

Magnetorheological biofluid and its applications

Sanjay Kumar, Bharat Kumar, Himanshu Shekhar Gupta, Ashish Kumar Sinha, Rakesh Sehgal, M. F. Wani, and Mukund Dutt Sharma

12.1 INTRODUCTION

Smart materials represent a new generation that outperforms traditional structural and functional materials. These materials can instantly alter their shape or size with minimal heat input or instantly transform from a liquid state to a semi-solid state when exposed to an external magnetic field. Smart materials exhibit adaptive responses to external stimuli, such as loads or environmental factors, showcasing inherent intelligence [1]. Intelligent materials include various types, such as ferrofluids, electrorheological fluids (ERFs), ferrogels, magneto-responsive elastomers (MREs), piezoelectric alloys (PEAs), and shape memory alloys (SMAs) [2]. Magnetorheological fluids (MRFs) are a type of magnetic fluid. These fluids are produced by embedding micro-sized ferrimagnetic particles, like carbonyl iron (CI), into nonmagnetic carrier liquids, such as silicone oil (SO), mineral oil (MO), or ionic solutions. MRFs exhibit a distinctive characteristic of reversibly transitioning their structure between liquid and semi-solid states in response to an external magnetic field. By applying such a field, rheological properties such as yield stress and viscosity can undergo significant alterations [3]. Due to their unique properties, MRFs find extensive applications across various sectors, including automotive (MR clutches and brakes), mechanical (dampers), construction (seismic dampers), and medical industries (drug delivery) [4–8].

One of the challenges with MRFs is the high density of CIPs, leading to a density mismatch between the dispersed and continuous phases, thereby reducing sedimentation stability. [9]. Moreover, the extensive utilization of existing MRFs is constrained by their elevated cost and substantial environmental hazards. To address this challenge, a mixing of biocompatible oil with some mineral oil is currently being utilized in the preparation of MRFs. The rheological characteristics of MRFs created through a blend of organic oils (specifically cottonseed and sunflower oils) with synthetic oil (silicone) were examined by Jinaga et al. [10]. Their research showed that MRFs blended with cottonseed oil demonstrated improved sedimentation stability. Shetty and Prasad [11] studied the magnetorheological properties of

DOI: 10.1201/9781003717362-12

magnetorheological fluid (MRF) based on nonedible vegetable oil. They noted that the yield stress exhibited a significant decrease once it reached 25 kPa. Kumar [12] prepared MRFs based on honey and examined their rheological properties.

Over time, researchers have explored various methods to improve sedimentation stability and MR performance. These methods include incorporating organic oils (rice bran, sunflower oil, cottonseed oil, and honey), surfactants, dimorphic particle mixtures, magnetic NPs, and core-shell particles and inorganic polymer-coated CIPs [13]. This chapter delves into the evolution of smart materials and the creation of biocompatible oil-based MRFs. Challenges related to these fluids are thoroughly examined alongside their applications across various sectors.

12.2 FIELD-RESPONSIVE FLUID

Field-responsive fluids are materials characterized by substantial alterations in their rheological characteristics upon exposure to an external field. These fluids are composed of a carrier liquid, which can be mineral oil or hydrocarbon oil, and solid particles. MRFs include particles that can be magnetized, whereas electrorheological fluids (ERFs) feature particles that react to electric fields. Ferrofluids, recognized as magnetic liquids, fall into the category of smart materials due to their ability to maintain liquid characteristics even when subjected to a magnetic field. Unlike other smart materials, ferrofluids generally do not exhibit particle chain formation or develop yield stress. However, they experience a body force that draws the fluid toward regions of intense magnetic fields. In summary, MRFs and ERFs both offer unique strengths and weaknesses, serving complementary rather than competing roles. Consequently, they find their niches in different markets and applications across various sectors [14, 15]. MRFs boast the ability to withstand higher levels of stress, whereas ER fluids stand out for their suitability in smaller systems.

12.2.1 Ferrofluid

Ferrofluids represent a distinct category of field-responsive fluids, setting them apart from traditional MRFs. They are created by dispersing magnetic NPs, size ranging from 5 to 15 nm and making up to 10% of the volume, into a nonmagnetic carrier oil, resulting in a stable colloidal mixture [16–18]. To form a stable colloidal suspension, superparamagnetic materials, such as iron oxides, magnesium-zinc (Mg-Zn) ferrites, and cobalt particles, are combined with a carrier fluid, which can be water, oil, or diester. The introduction of a magnetic field induces dipolar interactions among the particles, causing them to align along the direction of the magnetic field. In contrast to MRFs, ferrofluids do not exhibit yield stress; rather, they demonstrate viscosity that depends on the applied magnetic field [19]. MRFs, with their larger particles, exhibit distinct behavior compared to ferrofluids when subjected to the same magnetic field [20]. MRFs undergo a viscosity change and can solidify due to chain structure formation under a magnetic field. In contrast, ferrofluids typically maintain viscosity even under strong magnetic fields without forming chain structures and are more stable than MRFs since they are made with colloidal magnetic particles [16, 17].

12.2.2 MR fluid

MRFs are smart soft materials with diverse engineering applications due to their ability to alter viscoelastic properties under external magnetic fields. The discovery of MR fluid (MRF) is credited to Jacob Rabinow [21]. In the presence of an external magnetic field, these fluids undergo a rapid and reversible transformation from a fluidic to a semi-solid state, completing within milliseconds [22, 23]. In the absence of a magnetic field, the magnetic particles suspended in the carrier oil of magnetorheological (MR) fluids are dispersed haphazardly, making the MRFs act similarly to Newtonian fluids. However, the introduction of an external magnetic field triggers a dramatic change in the MRFs, transitioning them from a fluidic to a semi-solid state. This transformation is attributed to the magnetic dipole-dipole attraction among the particles, which aligns them into robust chain-like formations. The extraordinary capability of MRFs to shift between phases is depicted in Figure 12.1 [24, 25].

When exposed to a magnetic field, specific rheological characteristics of these fluids, like their apparent viscosity and yield stress, undergo swift alterations. Consequently, the fluid begins to display non-Newtonian behavior, characterized by a modest elevation in its viscosity [26, 27]. The viscosity of MRFs is influenced by various factors such as the characteristics of the carrier oil, the nature of the suspended particles, and the density of these particles within the fluid [6]. The viscosity of MRFs is affected by both the strength of the magnetic field and the rate of shear [28]. This characteristic

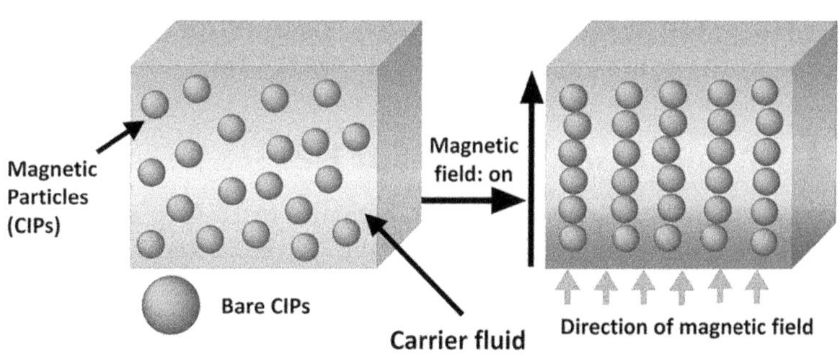

Figure 12.1 The schematic of the MRF working with and without magnetic field.

makes MRFs highly versatile and applicable across various industries, including automotive, aerospace, and civil engineering. In addition, MRFs have significant applications in the biomedical field, hyperthermia treatment, magnetic drug targeting, and cell separation via magnetic techniques [29–33].

12.2.3 Methods to improve sedimentation stability

The high surface area to volume ratio often decreases surface energy, causing particles to aggregate, resulting in significant challenges related to redispersibility and stability [9]. Numerous strategies have been proposed to mitigate settling rates and enhance the MR effect. This involves combining ferrofluid with MRF, coating cast iron and magnetite nanoparticles with lightweight materials such as wire or sheet-like particles, adding certain additives, using viscoelastic substances such as gels or other polymeric solutions, and forming thixotropic networks. This chapter discusses several recent techniques in progress, including additive-based MRFs, MRFs based on magnetite NPs, MRFs based on CI, and MRFs based on various carrier oils. In addition, conventional stabilization methods are explored for enhancing stabilization and the MR effect, as depicted in Figure 12.2.

12.2.4 Sedimentation stability improved by using additives

To improve sedimentation stability, additives, such as surfactants, polymer dispersants, carbon fiber, solid fillers, carbon nanotubes, silica nanoparticles, and clay, were added during the initial phase. The utilization of these additives led to the enhanced performance of MRFs, resulting in increased durability of MR devices.

The strategy to enhance the sedimentation stability in additive-enriched MRFs involves creating a gap between the magnetizable particles, thus inhibiting close interaction among them, as illustrated in Figure 12.3 [34]. Park et al. [35] explored the effects of nanoadditives (chromium dioxide) and grease on enhancing stability against gravitational forces. It was observed that fibril-like particles were more efficient than spherical particles in improving the sedimentation stability of MRFs [36].

12.2.5 MR fluid based on magnetic NPs

Iron oxides such as magnetite (Fe_3O_4), hematite (α-Fe_2O_3), and maghemite (γ-Fe_2O_3) are gaining significant interest due to their exceptional magnetic and biological characteristics [34]. Composite NPs based on magnetite, combined with various polymers such as PVB, PS, and conducting polymers, such as polypyrrole and PANI, are produced using diverse polymerization techniques. These hybrid additives are utilized in MRFs [37–39].

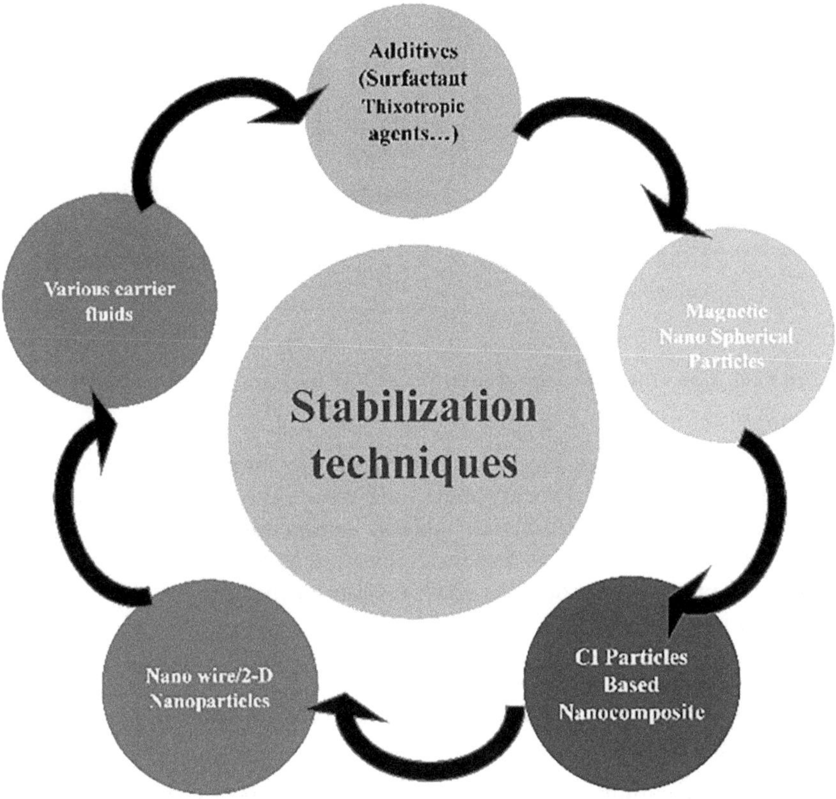

Figure 12.2 Various stabilization methods to enhance the sedimentation stability.

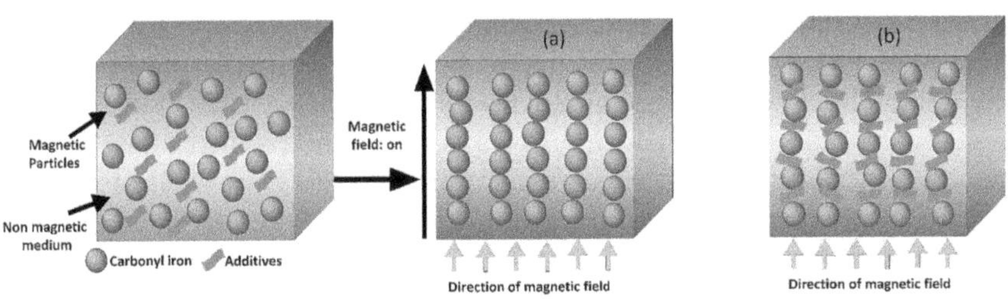

Figure 12.3 Schematic representation of MRFs (a) without additives (b) with additives.

The aforementioned studies demonstrate that solvent evaporation emerges as the optimal method for fabricating magnetite-based composite NPs [38].

Suspension polymerization or in situ dispersion polymerization techniques were used as an alternative for the preparation of spherical nanocomposite particles. The properties of NPs are governed by two primary

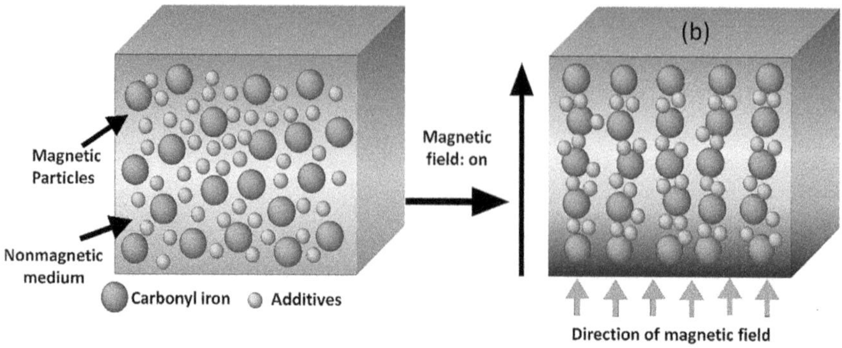

Figure 12.4 The schematic representation of MRF by incorporating magnetic NPs.

factors: their size and surface characteristics [40]. Composite NPs, including iron oxide, polymer, hydrocarbonate, and their derivatives, are utilized as additives in MR suspensions to address sedimentation problems. These compounds are examined as dispersed phases in MR suspensions because of their attributes, such as density, quick response time, and low surface area, as shown in Figure 12.4 [37, 41–43].

12.2.6 Core-shell/nanocomposite particles-based MR fluids

An ideal MRF should maintain low viscosity for easy flow within the components, and adopting core-shell CIPs with an appropriate coating layer has proven an effective method for solving sedimentation issues. This coating enhances suspension stability in magnetic suspensions by lowering particle density and improving particle durability [44]. Cho et al. [45] studied the impact of PMMA (produced through dispersion polymerization) coating on CIPs. The researchers reported that encapsulating CIPs with PMMA reduced particle density and improved suspension stability, albeit with a minor reduction in yield stress. Furthermore, the polymer shell of PMMA effectively prevented CIP agglomeration and shielded the core from external degradation effects.

Microparticles of CI are processed to enhance compatibility between the CI surface and the monomer [46]. This section delves into the current stabilization methods involving incorporating wire-shaped, 2D sheet-like structures, and flake-like particles as additives in MRFs. Nanoparticle-based MRFs exhibit lower stability than MRFs containing nanowires or nanosheets/flake-shaped particles, attributed to their inferior consistency. MRFs incorporating elongated particles experience reduced sedimentation [47].

12.2.7 Various carrier oil-based MR fluids

Several researchers have suggested a contemporary approach to address sedimentation problems in MRFs by employing higher-viscosity fluids such as lubricating oil or gel. The next generation of MRFs can be developed by dispersing magnetic particles into a gel, which effectively reduces particle sedimentation. Enhancing suspension stability can be achieved by applying polymeric gel onto the surface of CIPs [48]. However, a significant challenge arises from an unwanted increase in viscosity without a magnetic field, with PEO being one of the most commonly used polymers for creating polymeric solutions. Its solubility in various solvents and low toxicity make it well-suited for coating applications [49, 50]. Kim et al. [51] prepared an MR suspension by using a polymeric suspension incorporating bare CIPs. The authors reported that the prepared MR suspension has higher viscosity and yield stress compared to the conventional MRF.

12.3 APPLICATION

In the early 1990s, MRFs were successfully commercialized for use in aerobic exercise equipment rotary brakes. This marked a significant milestone in the advancement and practical application of MRFs in real-world scenarios [2]. MRFs are highly valued for applications that need adjustable fluid flow, such as in brakes, seismic dampers, vibration control for bridges, shock absorbers, machine and washing machine dampers, polishing materials, and active vibration control systems, due to their magnetic field responsiveness [2, 52]. MRFs are also applied in diverse fields such as chemical sensing, sound control, thermal regulation, electro-optic devices (including magnetic sensors and light valves), MR orifices, hydraulic valves, machining aids, rotating seals, robotics, and beyond [52, 53]. MRF commercialization has experienced swift and significant growth in recent years, as detailed in the subsequent subsections.

12.3.1 Mechanical engineering innovations and automotive implementations

MRF dampers are utilized in vibration control systems to reduce shock and counter movement, benefiting from their significant damping force, low power consumption, and other advantages [8, 54]. An MRF damper was introduced for trucks and vehicles, specifically for the seat's suspension, in 1998, as depicted in Figure 12.5. The Cadillac Seville STS, introduced in January 2002, featured a high-tech suspension system [23]. MR dampers offer distinct properties that are highly beneficial for applications such as vibration control systems, washing machines, seismic mitigation in structural engineering,

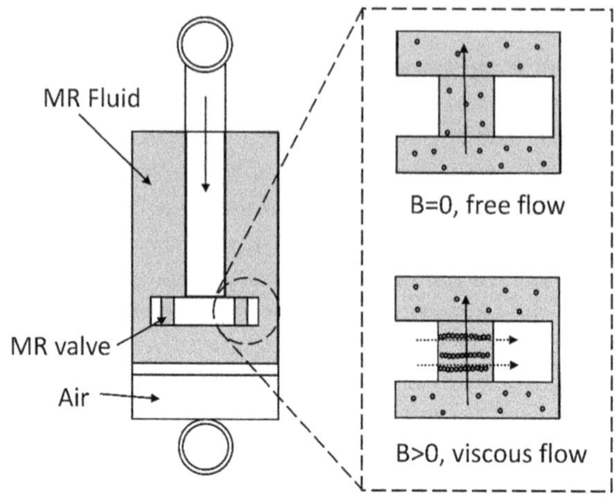

Figure 12.5 Schematic of MR damper with field off and on condition.

prosthetic devices, and mitigating wind-induced vibrations in cable-stayed bridges [46]. A new self-power damper for washing machines was introduced, incorporating energy-harvesting tech to directly transfer power from induction coils to the damper's excitation coils for damping force [55].

MRFs find application in automotive technology, notably in brake and clutch systems, operating primarily in shear mode but adaptable to flow mode. Lord Corporation's MRF rotary brake, for instance, offers controllable rotary resistance with compactness, smooth operation, and low power consumption [8].

12.3.2 Aerospace applications

The MRF damper enhances safety in helicopter seats during crashes, reducing impact on the pilot's spine. In addition, MREAs offer an innovative method for effectively tailoring energy absorption profiles. The MREA can effectively adjust the load profile to match the seat shock during a crash by employing feedback control. Among seat suspensions, there are three categories: passive, semi-active, and active. The semi-active method is particularly favored due to its lower power requirements compared to the active method and its lack of stability issues. Many researchers have been motivated to develop innovative seat suspensions that offer improved shock absorption and vibration attenuation [56].

12.4 CONCLUSION

This chapter focuses on MRFs and their stabilization techniques. A significant issue in MR technology is the sedimentation stability and the high cost associated with these fluids. In addition, the oxidation of magnetic particles poses a challenge for the widespread use of these fluids. This chapter delves into detailed solutions for sedimentation stability and summarizes the applications of these fluids. Hybrid core-shell particles are utilized to address sedimentation stability and particle oxidation, paving the way for future research in this area for upcoming researchers.

REFERENCES

[1] S. L. Vieira, C. Ciocanel, P. Kulkarni, A. Agrawal, and N. Naganathan, "Behaviour of MR fluids in squeeze mode," *Int. J. Veh. Des.*, vol. 33, no. 1–3, pp. 36–49, 2003.

[2] M. Kciuk and R. Turczyn, "Properties and application of magnetorheological fluids," *J. Achiev. Mater. Manuf. Eng.*, vol. 18, no. 1–2, pp. 127–130, 2006.

[3] J. D. Carlson and M. R. Jolly, "MR fluid, foam and elastomer devices," *Mechatronics*, vol. 10, no. 4–5, pp. 555–569, 2000.

[4] P. P. Phule, "Magnetorheological (MR) fluids: principles and applications," *Smart Mater. Bull.*, vol. 2001, no. 2, pp. 7–10, 2001.

[5] O. Ashour, C. A. Rogers, and W. Kordonsky, "Magnetorheological fluids: materials, characterization, and devices," *J. Intell. Mater. Syst. Struct.*, vol. 7, no. 2, pp. 123–130, 1996.

[6] J. D. Carlson, D. M. Catanzarite, and K. A. St. Clair, "Commercial magnetorheological fluid devices," *Int. J. Mod. Phys. B*, vol. 10, no. 23–24, pp. 2857–2865, 1996.

[7] S. P. Kelso, "Experimental characterization of commercially practical magnetorheological fluid damper technology," in *Smart Structures and Materials 2001: Industrial and Commercial Applications of Smart Structures Technologies*, 2001, vol. 4332, pp. 292–299.

[8] J. Wang and G. Meng, "Magnetorheological fluid devices: principles, characteristics and applications in mechanical engineering," *Proc. Inst. Mech. Eng. Part L J. Mater. Des. Appl.*, vol. 215, no. 3, pp. 165–174, 2001.

[9] J. De Vicente, D. J. Klingenberg, and R. Hidalgo-Alvarez, "Magnetorheological fluids: a review," *Soft Matter*, vol. 7, no. 8, pp. 3701–3710, 2011, doi: 10.1039/c0sm01221a

[10] R. Jinaga, T. Jagadeesha, S. Kolekar, and S. B. Choi, "The synthesis of organic oils blended magnetorheological fluids with the field-dependent material characterization," *Int. J. Mol. Sci.*, vol. 20, no. 22, 2019, doi: 10.3390/ijms20225766

[11] B. G. Shetty and P. S. S. Prasad, "Rheological properties of a honge oil-based magnetorheological fluid used as carrier liquid.," *Def. Sci. J.*, vol. 61, no. 6, 2011.

[12] D. Kumar, "Experimental evaluation of honey as carrier fluid for MR brake application," In *Proceedings of the International Conference on Innovative Development in Science, Technology and Development (IDSTM)*, Desalpur, India, 2018.

[13] S. Kumar, R. Sehgal, M. F. Wani, and M. D. Sharma, "Stabilization and tribological properties of magnetorheological (MR) fluids: a review," *J. Magn. Magn. Mater.*, vol. 538, no. April, p. 168295, 2021, doi: 10.1016/j.jmmm.2021.168295

[14] D. J. Klingenberg, "Magnetorheology: applications and challenges," *Am. Inst. Chem. Eng. AIChE J.*, vol. 47, no. 2, p. 246, 2001.

[15] J.-H. Park and O.-O. Park, "Electrorheology and magnetorheology," *Korea-Aust. Rheol. J.*, vol. 13, no. 1, pp. 13–17, 2001.
[16] C. Scherer and A. M. Figueiredo Neto, "Ferrofluids: properties and applications," *Braz. J. Phys.*, vol. 35, pp. 718–727, 2005.
[17] J. Popplewell, "Technological applications of ferrofluids," *Phys. Technol.*, vol. 15, no. 3, p. 150, 1984.
[18] K. Butter, P. H. Bomans, P. M. Frederik, G. J. Vroege, and A. P. Philipse, "Direct observation of dipolar chains in ferrofluids in zero field using cryogenic electron microscopy," *J. Phys. Condens. Matter*, vol. 15, no. 15, p. S1451, 2003.
[19] J. Liu et al., "Field-induced structures in ferrofluid emulsions," *Phys. Rev. Lett.*, vol. 74, no. 14, p. 2828, 1995.
[20] E. Lemaire, Y. Grasselli, and G. Bossis, "Field induced structure in magneto and electro-rheological fluids," *J. Phys. II*, vol. 2, no. 3, pp. 359–369, 1992.
[21] Ubaidillah, J. Sutrisno, A. Purwanto, and S. A. Mazlan, "Recent progress on magnetorheological solids: materials, fabrication, testing, and applications," *Adv. Eng. Mater.*, vol. 17, no. 5, pp. 563–597, 2015, doi: 10.1002/adem.201400258
[22] M. T. López-López, P. Kuzhir, S. Lacis, G. Bossis, F. González-Caballero, and J. D. G. Durán, "Magnetorheology for suspensions of solid particles dispersed in ferrofluids," *J. Phys. Condens. Matter*, vol. 18, no. 38, 2006, doi: 10.1088/0953-8984/18/38/S18
[23] C. Rinaldi, A. Chaves, S. Elborai, X. He, and M. Zahn, "Magnetic fluid rheology and flows," *Curr. Opin. Colloid Interface Sci.*, vol. 10, no. 3–4, pp. 141–157, 2005, doi: 10.1016/j.cocis.2005.07.004
[24] J. Rabinow, "The magnetic fluid clutch," *Trans. Am. Inst. Electr. Eng.*, vol. 67, no. December, pp. 1308–1315, 1948, doi: 10.1109/T-AIEE.1948.5059821
[25] L. Rodríguez-Arco, M. T. López-López, P. Kuzhir, G. Bossis, and J. D. G. Durán, "Optimizing the magnetic response of suspensions by tailoring the spatial distribution of the particle magnetic material," *ACS Appl. Mater. Interfaces*, vol. 5, no. 22, pp. 12143–12147, 2013, doi: 10.1021/am404035w
[26] Y. Rabbani, M. Ashtiani, and S. H. Hashemabadi, "An experimental study on the effects of temperature and magnetic field strength on the magnetorheological fluid stability and MR effect," *Soft Matter*, vol. 11, no. 22, pp. 4453–4460, 2015, doi: 10.1039/c5sm00625b
[27] G. R. Iglesias, M. T. López-López, J. D. G. Durán, F. González-Caballero, and A. V. Delgado, "Dynamic characterization of extremely bidisperse magnetorheological fluids," *J. Colloid Interface Sci.*, vol. 377, no. 1, pp. 153–159, 2012, doi: 10.1016/j.jcis.2012.03.077
[28] X. Wang and F. Gordaninejad, "Study of magnetorheological fluids at high shear rates," *Rheol. Acta*, vol. 45, no. 6, pp. 899–908, 2006, doi: 10.1007/s00397-005-0058-y
[29] D. Case, B. Taheri, and E. Richer, "Design and characterization of a small-scale magnetorheological damper for tremor suppression," *IEEE/ASME Trans. Mechatron.*, vol. 18, no. 1, pp. 96–103, 2011.
[30] B. Gonenc and H. Gurocak, "Virtual needle insertion with haptic feedback using a hybrid actuator with DC servomotor and MR-brake with Hall-effect sensor," *Mechatronics*, vol. 22, no. 8, pp. 1161–1176, 2012.
[31] M. Sedlacik et al., "Correlation of structural and magnetic properties of Fe3O4 NPs with their calorimetric and magnetorheological performance," *J. Magn. Magn. Mater.*, vol. 326, pp. 7–13, 2013.

[32] G. A. Flores, R. Sheng, and J. Liu, "Medical applications of magnetorheological fluids a possible new cancer therapy," *J. Intell. Mater. Syst. Struct.*, vol. 10, no. 9, pp. 708–713, 1999.
[33] G. M. Kamath, N. M. Wereley, and M. R. Jolly, "Characterization of magnetorheological helicopter lag dampers," *J. Am. Helicopter Soc.*, vol. 44, no. 3, pp. 234–248, 1999.
[34] B. J. Park, F. F. Fang, and H. J. Choi, "Magnetorheology: materials and application," *Soft Matter*, vol. 6, no. 21, pp. 5246–5253, 2010, doi: 10.1039/c0sm00014k
[35] J. H. Park, M. H. Kwon, and O. O. Park, "Rheological properties and stability of magnetorheological fluids using viscoelastic medium and nanoadditives," *Korean J. Chem. Eng.*, vol. 18, no. 5, pp. 580–585, 2001, doi: 10.1007/BF02706371
[36] F. F. Fang, I. B. Jang, and H. J. Choi, "Single-walled carbon nanotube added carbonyl iron suspension and its magnetorheology," *Diam. Relat. Mater.*, vol. 16, no. 4–7 SPEC. ISS., pp. 1167–1169, 2007, doi: 10.1016/j.diamond.2006.11.069
[37] M. K. Hong, B. J. Park, and H. J. Choi, "Preparation and physical characterization of polyacrylamide coated magnetite particles," *Phys. Status Solidi Appl. Mater. Sci.*, vol. 204, no. 12, pp. 4182–4185, 2007, doi: 10.1002/pssa.200777350
[38] B. J. Park, J. L. You, H. J. Choi, S. Y. Park, and B. Y. Lee, "Synthesis and magnetorheological characterization of magnetite nanoparticle and poly(vinyl butyral) composite," *IEEE Trans. Magn.*, 2009, vol. 45, no. 6, pp. 2460–2463, doi: 10.1109/TMAG.2009.2018612
[39] F. F. Fang, H. J. Choi, and W. S. Choi, "Two-layer coating with polymer and carbon nanotube on magnetic carbonyl iron particle and its magnetorheology," *Colloid Polym. Sci.*, vol. 288, no. 3, pp. 359–363, 2010, doi: 10.1007/s00396-009-2164-3
[40] C. Burda, X. Chen, R. Narayanan, and M. A. El-Sayed, "Chemistry and properties of nanocrystals of different shapes," *Chem. Rev.*, vol. 105, no. 4, pp. 1025–1102, 2005, doi: 10.1021/cr030063a
[41] Y. D. Liu, H. J. Choi, and S. B. Choi, "Controllable fabrication of silica encapsulated soft magnetic microspheres with enhanced oxidation-resistance and their rheology under magnetic field," *Colloids Surfaces A Physicochem. Eng. Asp.*, vol. 403, pp. 133–138, 2012, doi: 10.1016/j.colsurfa.2012.04.002
[42] S. W. Ko, J. Y. Lim, B. J. Park, M. S. Yang, and H. J. Choi, "Magnetorheological carbonyl iron particles doubly wrapped with polymer and carbon nanotube," *J. Appl. Phys.*, vol. 105, no. 7, pp. 103–106, 2009, doi: 10.1063/1.3058674
[43] Y. H. Kim, B. J. Park, H. J. Choi, and Y. Seo, "Coating of magnetic particle with polystyrene and its magnetorheological characterization," *Phys. Status Solidi Appl. Mater. Sci.*, vol. 204, no. 12, pp. 4178–4181, 2007, doi: 10.1002/pssa.200777353
[44] W. H. Chuah, W. L. Zhang, H. J. Choi, and Y. Seo, "Magnetorheology of Core-shell structured carbonyl iron/polystyrene foam microparticles suspension with enhanced stability," *Macromolecules*, vol. 48, no. 19, pp. 7311–7319, 2015, doi: 10.1021/acs.macromol.5b01430
[45] M. S. Cho, S. T. Lim, I. B. Jang, H. J. Choi, and M. S. Jhon, "Encapsulation of spherical iron-particle with PMMA and its magnetorheological particles," *IEEE Trans. Magn.*, vol. 40, no. 4 II, pp. 3036–3038, 2004, doi: 10.1109/TMAG.2004.830413

[46] M. Ashtiani, S. H. Hashemabadi, and A. Ghaffari, "A review on the magnetorheological fluid preparation and stabilization," *J. Magn. Magn. Mater.*, vol. 374, pp. 716–730, 2015.
[47] R. C. Bell, J. O. Karli, A. N. Vavreck, D. T. Zimmerman, G. T. Ngatu, and N. M. Wereley, "Magnetorheology of submicron diameter iron microwires dispersed in silicone oil," *Smart Mater. Struct.*, vol. 17, no. 1, p. 15028, 2008, doi: 10.1088/0964-1726/17/01/015028
[48] B. Wei, X. Gong, W. Jiang, L. Qin, and Y. Fan, "Study on the properties of magnetorheological gel based on polyurethane," *J. Appl. Polym. Sci.*, vol. 118, no. 5, pp. 2765–2771, 2010, doi: 10.1002/app.32688
[49] E. A. Van Nierop, D. J. Keupp, and H. A. Stone, "Formation of free films of aqueous solutions of poly(ethylene oxide): the influence of surfactant," *Europhys. Lett.*, vol. 88, no. 6, p. 66005, 2009, doi: 10.1209/0295-5075/88/66005
[50] S. T. Lim, C. H. Hong, H. J. Choi, P. Y. Lai, and C. K. Chan, "Polymer turbulent drag reduction near the theta point," *Epl*, vol. 80, no. 5, p. 58003, 2007, doi: 10.1209/0295-5075/80/58003
[51] M. S. Kim, Y. D. Liu, B. J. Park, C. Y. You, and H. J. Choi, "Carbonyl iron particles dispersed in a polymer solution and their rheological characteristics under applied magnetic field," *J. Ind. Eng. Chem.*, vol. 18, no. 2, pp. 664–667, 2012, doi: 10.1016/j.jiec.2011.11.062
[52] H. Lu *et al.*, "The roles of magnetorheological fluid in modern precision machining field: a review," *Front. Mater.*, vol. 8, p. 678882, 2021.
[53] D. Susan-Resiga and P. Barvinschi, "Nano-microstructured magnetorheological fluids and engineering applications," *Rom. J. Tech. Sci. Appl. Mech.*, vol. 65, no. 2, pp. 89–122, 2020.
[54] A. Hajalilou, S. Amri Mazlan, H. Lavvafi, and K. Shameli, *Field Responsive Fluids as Smart Materials*. Springer, 2016.
[55] D. Hua, X. Liu, Z. Li, P. Fracz, A. Hnydiuk-Stefan, and Z. Li, "A review on structural configurations of magnetorheological fluid based devices reported in 2018–2020," *Front. Mater.*, vol. 8, p. 640102, 2021.
[56] G. J. Hiemenz, Y.-T. Choi, and N. M. Wereley, "Semi-active control of vertical stroking helicopter crew seat for enhanced crashworthiness," *J. Aircr.*, vol. 44, no. 3, pp. 1031–1034, 2007.

Chapter 13

Development, future scope, and challenges in biotribology

M. Natesh, D. Amrishraj, and J. Anita Jessie

13.1 DEVELOPMENT IN BIOTRIBOLOGY

13.1.1 Joint tribology

13.1.1.1 Natural synovial joints

Understanding tribology in natural synovial joints and artificial joint replacements requires a thorough investigation into lubrication mechanisms, wear patterns, and the influence of biomaterials on joint performance. This overview presents important findings from various studies in these areas [1]. A thorough examination was conducted on natural synovial joints, which are known for their fluid and cartilage composition, as well as their association with conditions such as osteoarthritis [2]. Various studies examined the characteristics of both healthy and diseased joints, along with different materials used for replacements, like hydrogel and ultra-high molecular weight polyethylene (UHMWPE) [3–5]. Explorations were conducted on the mechanics of lubrication in natural joints, with a specific emphasis on the surfaces of articular cartilage and the lubricating qualities of synovial fluid. However, certain studies failed to consider the dynamic nature of chondrocyte cells in articular cartilage by using simplified loading conditions [6]. The lubrication process in natural synovial joints is a highly intricate and dynamic interplay between various factors, such as phosphatidylcholine lipids and hyaluronic acid. Previous studies primarily focused on friction measurements, overlooking the examination of wear mechanisms and cartilage damage. Addressing these concerns, novel setups were utilized to visualize lubricating films and measure friction coefficients [7]. While there was a strong focus on experimental studies, analytical and computational models also became increasingly popular. Various models were used to investigate the contact between cartilage surfaces, including theoretical lubrication models and atomistic models [8, 9]. The focus was on boundary lubrication mechanisms, specifically phosphatidylcholine lipids and hyaluronic acid. Various studies have also delved into interventions aimed at restoring lubrication in cartilage that has undergone degradation [10–12].

DOI: 10.1201/9781003717362-13

The study encompassed tissue engineering, examining the tribological aspects of tissue-engineered cartilages and exploring alternative materials such as hydrogels. The evaluations of biomaterials showed different levels of damage to the cartilage during hemi-arthroplasty. New hydrogels inspired by nature that imitate the properties of biphasic cartilage, along with innovative biomaterials that have improved lubrication, were presented [13]. Artificial joint replacements involve extensive studies focused on enhancing biomaterials and refining implant designs. Wear studies, including computational modeling, offered valuable insights. Insights into surgical cup placement effects and early detection of hip implant failures were gained through in vivo measurements using instrumented prostheses with sensors [14]. This interdisciplinary study plays a crucial role in pushing forward the field of orthopedics and biomechanics. It has a profound impact on enhancing the design and refinement of materials used in synovial joints and joint replacements [15].

Artificial joints play a vital role in the treatment of joint diseases, yet they face challenges such as loosening of prosthetic components caused by reactions triggered by wear debris. Investigating tribological studies is crucial in order to enhance the performance of artificial joints, specifically by analyzing friction, lubrication, and wear. Wear debris and corrosion can result in negative biological responses, which can lead to clinical complications. It is crucial to consider patient-specific factors and variations in anatomy, loading, and surgical techniques when integrating tribological studies with biological and clinical investigations [16]. Various biomaterials, such as ultra-high molecular weight polyethylene (UHMWPE) and ceramics, are used to reduce wear and enhance stability. Extensive studies have indicated that the wear reduction of highly cross-linked UHMWPE, particularly when used with ceramics, holds great potential. Ceramic-on-ceramic combinations show very few signs of wear. Although there are difficulties associated with metal-on-metal bearings, it is essential to have a comprehensive understanding of design parameters, surgical techniques, and patient-specific variables in order to improve the longevity and performance of artificial joints in clinical settings [17].

13.1.2 Skin tribology

13.1.2.1 Friction behavior of human skin

The behavior of human skin during mechanical contact is quite complex, resembling that of soft elastomers. It is a multilayered composite material with unique properties. Conventional friction rules are not appropriate, which has resulted in the utilization of elastomeric theories. Under dry conditions, the skin's properties of non-homogeneity, nonlinearity, and viscoelasticity are taken into account in a two-term friction model that focuses on adhesion and deformation. The skin-material interface is primarily influenced by attractive forces, which contribute significantly to adhesion. Deformation mechanisms, on the other hand, have a secondary impact [18]. The skin's friction is affected by various factors, including its properties, the characteristics of the materials it comes into contact with, and any intermediate layers such as cosmetics or sweat. Various factors can influence skin friction, including the type of material, physiological conditions, and mechanical parameters, such as normal load, sliding velocity, age, gender, ethnicity, and anatomical region. Grasping these complexities is essential for explaining the tribological behavior of human skin. Figure 13.1 provides a visual representation of the key factors that impact the friction behavior of human skin [19].

13.1.2.2 Surface properties of human skin

Scientists have been studying the relationship between the texture of human skin and how it affects friction. Skin roughness can vary across different parts of the body, ranging from 10 to 200 μm. Various studies have been conducted to measure skin roughness using parameters such as average roughness (Ra). These studies aim to quantify the roughness of the skin, with Ra values typically ranging from 10 to 200 μm. Nevertheless, the connection between skin topography and friction is quite complex. Different studies have varying findings regarding the relationship between skin surface roughness (Ra) and friction coefficients [20]. Some research indicates that including Ra in multi-regression analyses, along with factors such as skin moisture, can improve predictability. The density of primary lines on the skin's surface, which is affected by age, has been found to have a direct relationship with increased

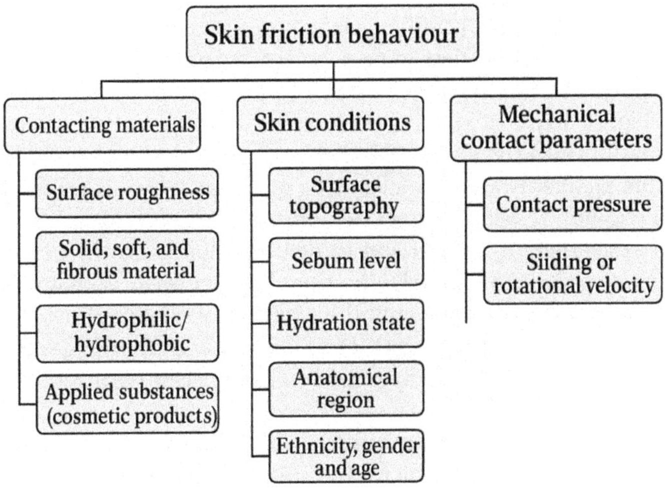

Figure 13.1 Important factors influencing the friction behavior of human skin.

friction. The understanding of the impact of sebum lipids on skin tribology remains incomplete, and the existing correlations between skin surface lipid content and friction suggest a limited influence. The level of skin hydration has a notable impact on friction, as increased moisture leads to higher skin surface friction. The study of skin tribology is essential in a wide range of applications, including cosmetics, medical and sports-related fields, textiles, and ergonomics. Skin friction is influenced by factors such as material properties and sliding velocity. Gaining a comprehensive understanding of these complex factors is crucial for uncovering the wide-ranging implications of skin friction in various fields [21].

13.1.3 Oral tribology

13.1.3.1 Wear of human teeth

The structure and texture of human teeth have developed over millions of years of evolution, creating a natural anti-wear system. Despite the intricate nature of the oral environment, human teeth effortlessly carry out their chewing function, demonstrating remarkable properties that are not widely comprehended [22]. Exploring the durability of human teeth reveals five key factors: a cleverly designed structure, surface treatments, a dense and organized microstructure, the ability to self-repair, and a one-of-a-kind lubrication system [23]. Human teeth are made up of different components, such as enamel, dentine-enamel junction, dentine, and pulp. The enamel, being the hardest tissue in the body, is primarily composed of inorganic materials, which are arranged in tightly-packed hydroxyapatite rods. Dentine, situated between enamel and pulp, is a hydrated biological composite that possesses a lower hardness but is known for its elasticity and softness. The dentine-enamel junction plays a vital role in biological processes. Wearing down of teeth, caused by chewing, grinding, clenching, and chemical factors, can lead to various clinical issues that impact both the appearance and function of the teeth [24]. The relationship between tooth movement and wear is illustrated in Figure 13.2.

Investigation into the sliding friction and wear characteristics of enamel and dentine against titanium balls, using artificial saliva as a lubricant, has yielded intriguing results. Enamel demonstrates superior friction reduction and wear resistance in comparison to dentine, thanks to the occurrence of microfracture processes and delamination. Dentine wear is characterized by the formation of ductile chips and the creation of strong plows. The tribological properties of natural teeth can vary depending on age. Permanent teeth in the young and middle ages tend to have better wear resistance compared to primary teeth [25].

The process of dental erosion is often worsened by factors, such as bruxism or aggressive brushing, which can cause mechanical wear. This leads to the chemical dissolution of hydroxyapatite crystals. The erosion can result in tooth sensitivity, changes in appearance, and issues with functionality.

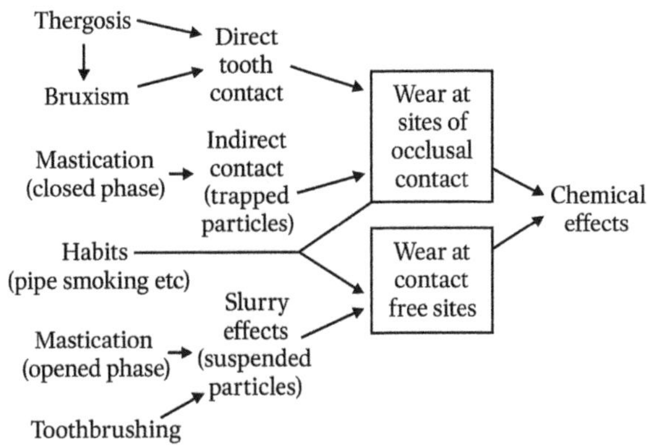

Figure 13.2 Relation between movement and wear of teeth.

Taking proactive steps can help reduce the risk of dental issues. It is important to limit contact with acidic substances, keep up with regular oral care, and consider using products that contain fluoride. Adjusting your diet and lifestyle can also contribute to minimizing erosion risk [26]. Factors such as the location, duration, and acidity of exposure have a notable impact on the rate of wear that occurs during dental erosion. The susceptibility of enamel to damage rises as the pH levels of erosive substances decrease. The interaction of chemical erosion and mechanical wear can lead to different wear patterns and rates depending on the applied loads [27].

Gaining a comprehensive understanding of these intricate interactions is vital for dental research and practice. Human teeth have a remarkable ability to repair themselves, thanks to the important inorganic ions found in saliva. This provides a certain level of protection. Nevertheless, certain factors such as bruxism, xerostomia, and elevated levels of tetracycline can contribute to significant tooth wear.

Tooth wear can be caused by bruxism, which involves heavy pressure on the teeth. Another factor that contributes to tooth wear is xerostomia, which is a condition characterized by reduced salivary flow. Brushing your teeth by itself has little effect on the wear of enamel and dentine, although dentine abrasion can happen if you use abrasive toothpaste. The intricate relationship between chemical erosion, mechanical abrasion, and a multitude of factors highlights the significance of taking preventive measures and practicing good oral hygiene to preserve dental health [28]. Figure 13.3 displays SEM micrographs of the nanoscratches found on the surface of human tooth enamel at 100 mN.

13.1.3.2 Wear of artificial dental materials

Dental implants often utilize titanium alloys, such as commercially pure titanium (CP-Ti), due to their excellent biocompatibility. Nevertheless, the tribological properties of these materials may be subpar, which can have a

Figure 13.3 SEM micrographs of the nanoscratches on the surface of human tooth enamel at 100 mN (a) the scratch, (b) wear particles, (c) the central area of scratch, (d) the unworn surface of enamel.

negative effect on their wear resistance. Enhancing these properties can be achieved by alloying CP-Ti. In orthodontics, there may be some discomfort due to the friction between brackets and the oral mucosa. Various elements can influence the level of friction in orthodontic systems, such as the materials used for arch wires and brackets, their dimensions, surface characteristics, cleanliness, and the presence of lubrication. An investigation into the wear patterns of titanium alloys in contact with human cortical bone has shed light on potential concerns regarding implant fixation. The study highlights the possibility of wear occurring at the bone-implant interface, which could pose challenges for long-term implant stability. Gaining insights into these tribological behaviors is crucial for creating dental implants and orthodontic appliances that are resilient and provide lasting comfort for patients [29–31].

13.1.3.3 Tribological properties of saliva

Saliva, which is mostly water along with electrolytes and proteins, has an important function in the mouth. It acts as a lubricant and creates a protective barrier between the teeth and the soft tissues in the mouth. Almost immediately after coming into contact with the surfaces inside the mouth, proteins in saliva attach themselves, forming a thin film that helps to lubricate. The pellicle has a unique structure, consisting of a thin inner layer and a thick, well-hydrated outer layer [32]. This structure helps to minimize friction between oral surfaces, which in turn helps to protect teeth from wear and prevents the development of lesions on the mucosa and tongue. The inner salivary film shows a notable reduction in friction and wear, displaying excellent adhesion to tooth surfaces. The effectiveness of lubrication is affected by the characteristics of the surface it is applied to, including its smoothness, energy, and chemical makeup. The salivary pellicle serves as a natural shield, safeguarding teeth against acid attacks. Even though dental friction and wear are bound to happen, human teeth, being a remarkable natural wearable system, can last for many years. Exploring the remarkable resistance to wear exhibited by teeth can provide valuable insights for improving the durability of engineering systems subjected to friction [33]. This is illustrated in Figure 13.4.

13.1.4 Tribology of other biological systems

Cardiovascular tribological concerns encompass artificial heart valves, intravascular stents, and the dynamics of blood flow. Extensive studies have been conducted on the wear at interfaces in heart valves to ensure safety and durability. Exploring various coatings such as TiO_x, films has become crucial in the quest to prevent thrombus formation on non-adhesive surfaces. Nevertheless, clinical use is still restricted by factors such as friction, wear, and adhesion [34]. Achieving smooth and soft hair is a constant endeavor, with a strong emphasis on minimizing friction and maximizing adhesion in hair care. A study delves into the relationship between the properties of friction and the structure of hair. Studies on friction at a larger scale have revealed interesting effects related to directionality, which can provide valuable insights for the development of hair-care products. However, there is a

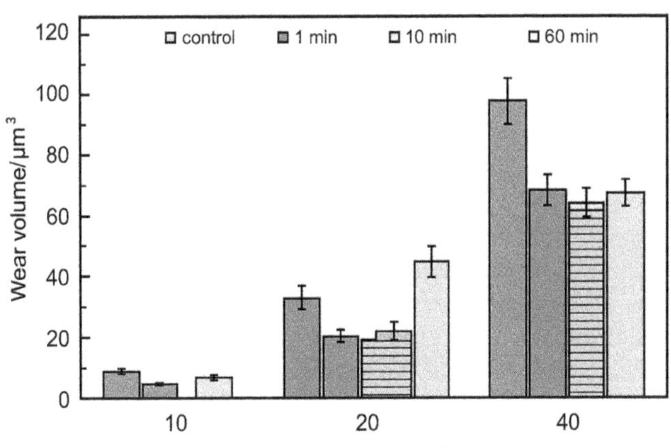

Figure 13.4 Wear volume of scratches on the enamel surfaces after different saliva adsorption times.

lack of research that focuses on important techniques, such as dietary habits, that can help prevent hair damage and improve its quality.

The eye is a well-moistened system essential for clear vision, as the tear film plays a vital role in safeguarding the cornea. Exploring ocular tribology, specifically in the context of contact lens usage, delves into the intricate dynamics between the eyelid, cornea, and lenses. Our investigation centers around tear film properties, contact lens adhesion, friction, and wearer comfort. Investigations examine the lubrication process that occurs during blinks and the forces of friction in saline solutions. However, delving deeper into the relationship between contact lens properties and biological responses, such as infection or tearing, requires additional research [35]. Geckos, tree frogs, and insects showcase fascinating skills in navigating smooth and inverted surfaces. Investigation delves into the mechanisms of adhesion, ascribing capillary and Vander Waals forces to gecko adhesion. Gecko toe detachment utilizes rolling and lever functions to quickly adjust adhesion and friction forces. There are ongoing difficulties in comprehending the connections between biological structures and frictional properties, as well as the theory of bionic design and manufacturing technology for these intricate attachment/detachment systems. Water striders have developed a remarkable adaptation that allows them to effortlessly glide across the water's surface. By utilizing their water-repellent legs, these incredible creatures are able to stay afloat and swiftly maneuver on top of the water. Interactions between animals and their environment in water, air, and soil reveal fascinating adaptations. Microscopic structures on fish scales help them swim more efficiently by reducing drag. Birds use herringbone riblet surfaces on their feathers to minimize drag. Animals that come into contact with soil, such as pangolins, have developed remarkable durability and a self-cleaning mechanism thanks to their unique corrugated scales. The attachment of fouling organisms is influenced by the microstructure of the surface of shells. Even with extensive research and attempts to replicate these structures using biomimetic methods, artificial surfaces still fall short of the remarkable efficiency found in natural bio-surfaces that have evolved over millions of years [36]. There is a lack of focus and understanding when it comes to plant tribology research, resulting in slow progress. Nevertheless, there has been a growing fascination with the lotus leaf and its remarkable ability to repel water, known as superhydrophobicity or the lotus effect. Lotus epidermal cells have unique structures that help minimize water contact. There is a growing interest in the bionic design of superhydrophobic surfaces, which draws inspiration from the lotus effect. These surfaces are finding applications in a wide range of fields. The study of algae tribology, specifically diatom microalgae, reveals their remarkable ability to self-lubricate as they grow. This discovery holds promise for developing highly effective lubrication techniques. Engineers are yet to explore the uncharted territory of unfamiliar plants that exhibit exceptional tribological behaviors, such as superlubricity [37].

13.1.5 Hip replacement overview

Surgeons carefully choose from a wide range of hip replacement brands available in the global market, considering patient characteristics and their own clinical expertise. The structure and materials of total hip replacements (THR) and resurfacing hip replacements (RHR) are depicted in Figure 13.5. By opting for RHRs, you can benefit from the enhanced stability provided by larger heads, while also preserving more bone. These options are exclusively available in MoM or CoM combinations. Plastic, metal, and ceramic are commonly used for various implant components [38].

13.1.6 Implant tribology

The passage covers different aspects of hip replacements, such as their geometry, materials, surface mechanics, friction, lubrication, and wear. When it comes to geometry, the size of the head (D_b) and the clearance (Cl) of hip implants play a vital role, regardless of whether they are normal or large head implants. Various materials are utilized for hip implants, such as plastics, metals, and ceramics, each possessing unique mechanical properties. Significant progress has been made in the field, with the introduction of highly cross-linked polyethylene (HXLPE) for plastic components, a range of metal alloys such as CoCrMo, and the development of advanced ceramic materials like Biolox delta [16].

The passage explores surface mechanics, highlighting the variations in pressure and contact width depending on the size and type of the implant. It delves into the topic of friction, showcasing the variations in coefficients among various bearing types. The intricate nature of lubrication is explored, focusing on the importance of maintaining a minimum film thickness and understanding different lubrication regimes. Wear, a significant clinical concern, is widely acknowledged as the leading factor behind hip implant

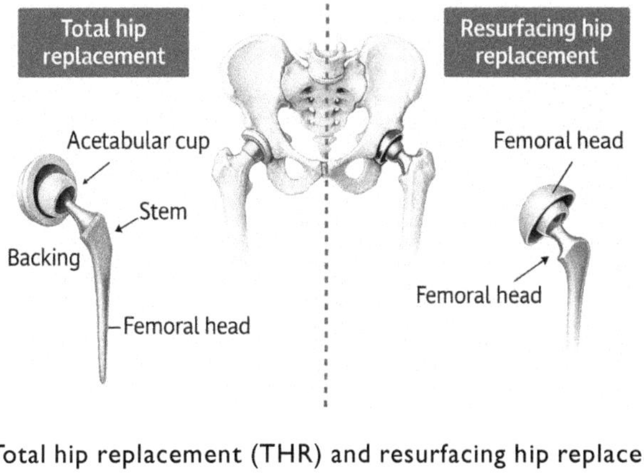

Figure 13.5 Total hip replacement (THR) and resurfacing hip replacement (RHR).

failure. The passage provides information on wear rates for various implant types and discusses the factors that can affect wear, including lubrication, geometry, and loading conditions. The text provides a comprehensive overview of hip replacement tribology, addressing the challenges in predicting wear coefficients and the impact of various materials and design considerations [39].

13.1.7 Tribological biomaterials in nature

Throughout the course of history, the intricate workings of nature's tribological systems have captivated the minds of both scientists and philosophers. Some interesting examples are gecko bristles that have remarkable adhesion due to certain forces, lotus leaves that are superhydrophobic and self-cleaning, shark skins that reduce surface drag with grooves, pangolin scales that resist wear and adhesion to soil, bamboo's hierarchical structure that ensures wear resistance, and articular cartilage's lubrication mechanism with water content. Bionics has a fascinating history, even though it is a relatively new concept in engineering materials. It involves simulating the structures and functions found in nature. Figure 13.6 displays various surfaces of tribological materials found in nature [40].

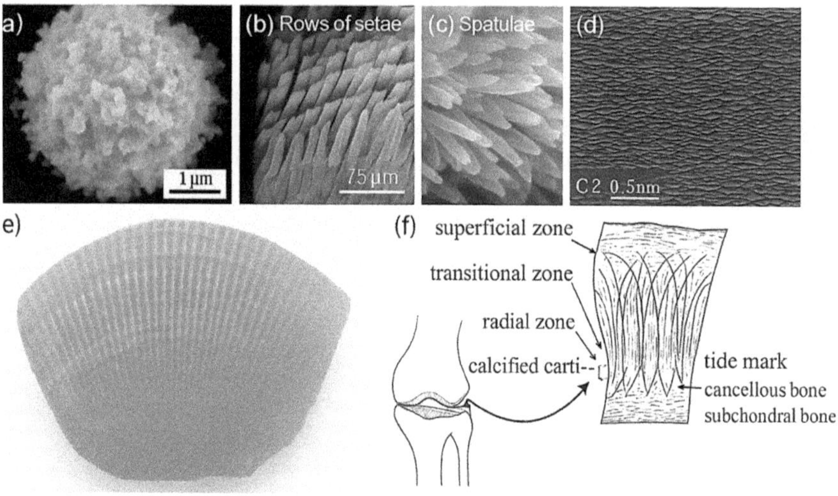

Figure 13.6 Several surfaces of tribological materials in nature: (a) SEM image of the surface of lotus leaf, SEM image of setal arrays (b) and spatulae (c) of geckos, (d) scale patterns of fast sharks, (e) photograph of pangolin scales, (f) Drawing shows the zonal anatomy of articular cartilage and collagen fiber orientation, and it can be seen the existence of hierarchical structure.

13.2 FUTURE SCOPE IN BIOTRIBOLOGY

13.2.1 Joint tribology

Considerable advancements have been achieved in comprehending the tribological mechanisms of natural synovial joints and utilizing this knowledge to improve the performance of artificial joints. Current efforts are centered around reducing wear and increasing the lifespan of prostheses. Early interventions, such as cartilage repair, are focused on prolonging the need for joint replacement. It is anticipated that tribological studies will have a greater significance in the coming years. Although there are multiple lubrication mechanisms at play in natural joints, gaining a complete understanding of them continues to be a complex task. Integrating biphasic cartilage deformation and boundary lubrication into models can be quite intricate [41].

13.2.2 Skin tribology

Anticipate exciting developments in the field of skin tribology, particularly in the realm of smart textiles and materials inspired by nature. These exciting advancements have the potential to enhance prosthetics, medical devices, and artificial skin by minimizing friction and enhancing comfort. The cosmetics industry stands to gain from the introduction of products that provide improved skin protection and increased user satisfaction. Wearable technology is poised for advancement as it embraces the ability to track and analyze various skin parameters. Future research will be influenced by environmental factors and sustainable practices, fostering interdisciplinary collaborations to drive innovations that enhance well-being [42].

13.2.3 Oral tribology

Exciting advancements are expected in the field of oral tribology, which will have a positive impact on dental care and oral health technologies. Future investigations will probably focus on innovative dental materials that have improved tribological properties, with the goal of enhancing the longevity of implants, crowns, and orthodontic appliances. Enhancing implant integration through bioactive coatings and surface modifications can promote faster healing. The advancements in 3D printing have the potential to completely transform the dental field by offering highly accurate and personalized prosthetics. Advanced dental tools, equipped with cutting-edge technology, can monitor oral hygiene and provide instant feedback. Collaborations between dental professionals, scientists, and engineers will be instrumental in driving these advancements, resulting in improved dental treatments and enhanced patient comfort [43, 44].

13.2.4 Tribology of other biological systems

Advancements in biological tribology are expected to be driven by collaborations across different disciplines, as well as the development of new materials and engineering solutions. Scientists will investigate natural lubrication systems, such as synovial joints, skin, and eyes, with the goal of understanding intricate mechanisms and developing biomimetic solutions. The incorporation of nanotechnology is anticipated to play a crucial role, as nanoscale coatings and lubricants imitate natural processes to enhance protection and minimize damage caused by friction [45]. The nanoscale approach has the potential to revolutionize various fields, including medical applications, wearable technology, and bio-inspired robotics. In addition, delving deeper into the study of the tribological properties of different biological systems, such as animals and plants, has the potential to spark creativity in the development of new materials. In addition, there might be an emphasis on creating lubricants that are sustainable and environmentally friendly, made from renewable sources. This would help minimize the impact on the environment and encourage sustainability in tribological applications. The future developments in biological tribology will be driven by collaborative efforts among biologists, materials scientists, engineers, and medical professionals. This multidisciplinary approach has the potential to bring about groundbreaking innovations [46].

13.2.5 Hip replacement overview

Hip replacement surgery involves the replacement of a damaged hip joint with a prosthetic implant, which effectively alleviates pain and enhances mobility. Implants such as stems, femoral heads, and acetabular cups are available in different materials, including metal-on-plastic (MoP), ceramic-on-ceramic (CoC), or metal-on-metal (MoM) [47]. Surgeons make their decisions based on the specific needs of each patient. Technological advancements such as highly cross-linked polyethylene (HXLPE), cobalt-chromium-molybdenum (CoCrMo), and ceramics have been developed to enhance wear resistance and durability. These innovations are aimed at improving the lifespan of implants and ultimately benefiting patients [16].

13.2.6 Cardiovascular physiology related to biotribology

13.2.6.1 Overall structure of blood vessels: Arteries

Arteries are composed of three layers. The outer layer, known as the tunica externa, connects the vessel to the surrounding tissue with collagen and elastic fibers. The middle layer, called the tunica media, contains concentric smooth muscles that enable the contraction of the vessel. Finally, the inner layer, known as the tunica interna, consists of elastic fibers, connective tissue, and a layer of endothelial cells. The endothelium has a vital role in establishing direct contact with medical devices for tribological interactions [48]. Figure 13.7 presents a schematic cross-sectional view of the structure of a blood vessel and endothelial cells.

13.2.6.2 Inner lining of blood vessels: The endothelium

The endothelium plays a crucial role in facilitating cellular exchange and maintaining proper blood flow. The monolayer of cells in the body has the ability to detect and respond to flow features, which in turn affects the vascular tone. These cells release certain factors that can either constrict or relax blood vessels [49]. This control governs the resistance of blood, which in turn regulates the supply of blood to organs. In addition, the endothelium plays a crucial role in maintaining the fluidity of blood, preventing clotting, and reducing inflammation. Damage can alter its function, which may have an impact on clotting and affect procedures such as stenting. This can hinder the process of re-endothelialization and increase the risk of stent thrombosis [50].

13.2.6.3 Tribology and endothelial cells: Friction-induced effects on cell health and behavior

The interactions between different surfaces can have an impact on the nucleus and cytoskeleton of cells, even without any direct contact. Cells have the remarkable ability to detect and respond to external forces, which can have a profound effect on how genes are regulated and how cells behave [51]. Research on endothelial cells has revealed the impact of various forces, such as indentation, shear stress, and strain, on their behavior. Understanding these effects is essential for their proper functioning under blood pressure [52]. Endothelial cells thrive under high unidirectional shear

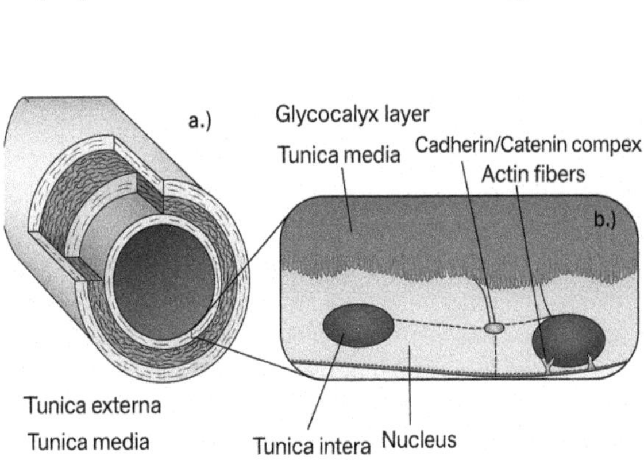

Figure 13.7 (a) Schematic cross-section view of blood vessel and (b) endothelial cell structure.

stress, which promotes their well-being. However, when exposed to low or oscillating stress, these cells can become inflamed and dysfunctional, potentially leading to disease. This suggests a possible connection between device friction and the development of certain health conditions. Friction has the potential to affect the functions of the cytoskeleton and genetics, which may lead to damage in cells [53].

13.2.6.4 Relevance and importance of vascular biotribology

This review explores the field of biotribology, specifically focusing on the study of endovascular medical devices. These devices include stent retrievers, thrombectomy tools, and advanced technologies, like the HARP robot, which are used within blood vessels. Highlighting the significance of stents and catheters in medical practice, these devices are designed to maintain the openness of blood vessels and fulfill diverse functions such as drug delivery. However, our current knowledge about the tribological interactions between these devices and the human body is still limited [54].

13.2.6.5 Studying vascular biotribology: Different approaches

Friction studies on medical device–tissue interactions can be conducted using different approaches, each with varying levels of complexity. These approaches include models that use artificial materials, in vitro studies conducted in a laboratory dish, ex vivo studies using animal tissue, in vivo studies conducted on live animals, and medical studies. Greater complexity leads to more accurate findings, but it also reduces the ability to replicate the results and demands more resources. Furthermore, in silico models are capable of simulating interactions computationally. The review section organizes experiments into different testing environments, including models, ex vivo, in vitro, in vivo, and in silico [55].

13.2.7 Biotribological model systems

13.2.7.1 Diatoms: Creators of glass castles

Diatoms are unicellular microalgae that have siliceous skeletons surrounded by thin organic cell walls. These skeletons form pillbox-like shells with various shapes and nanostructured frameworks. Discovered in various habitats, such as freshwater, marine environments, and damp surfaces, these organisms come in different sizes and species. They can range from planktonic to benthic forms, often forming chains [56]. Diatoms are highly useful in tribological studies and nanotechnology, as they can be used as templates for creating three-dimensional MEMS. Their silica surfaces are helpful for conducting investigations. Certain diatoms exhibit flexible components similar to rubber bands, while others possess robust underwater adhesives [57]. With their ease of cultivation and availability for microscopic examination, diatoms provide valuable insights into the field of tribological properties. Collaborations between different fields are investigating the possibilities in nanotechnology. Diatom shells show no signs of wear, making them ideal for nanotechnological applications [58]. Figures 13.8–13.10 displays scanning electron microscopy (SEM) images of the diatom species *Amphitetras antediluviana* Ehrenberg.

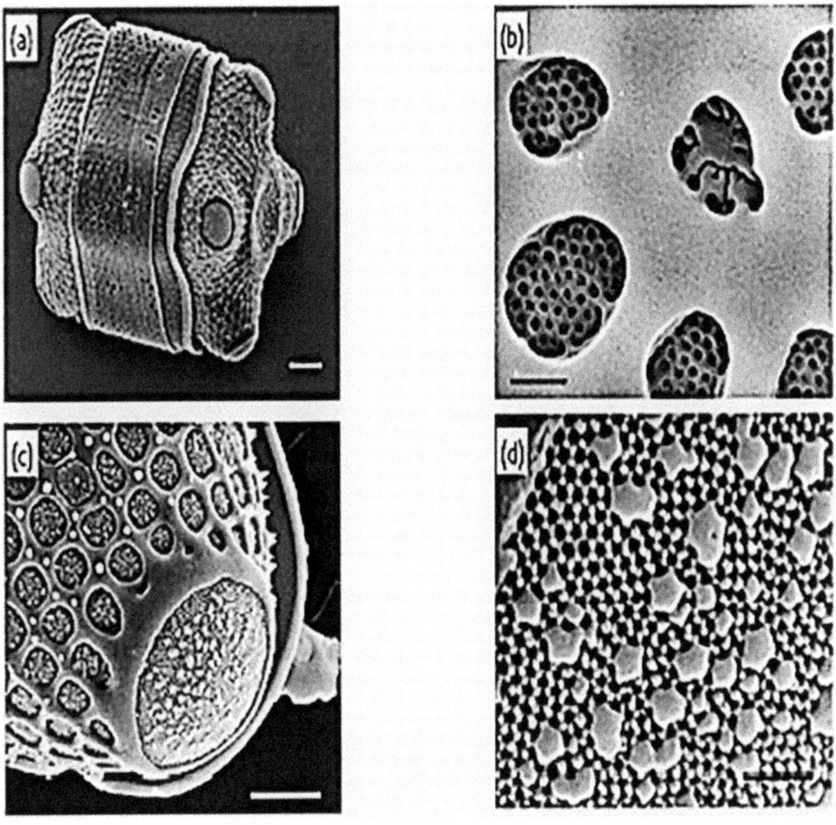

Figure 13.8 Scanning electron microscopy (SEM) images of the diatom species Amphitetrasantidiluvianum Ehrenburg. The sample was obtained from seaweed in Point Dume State Park, California. The whole cell can be seen in (a). The other images are close-ups of this cell. Scale bars: 20 μm (a), 1 μm (b), 5 μm (c), and 1 μm (d).

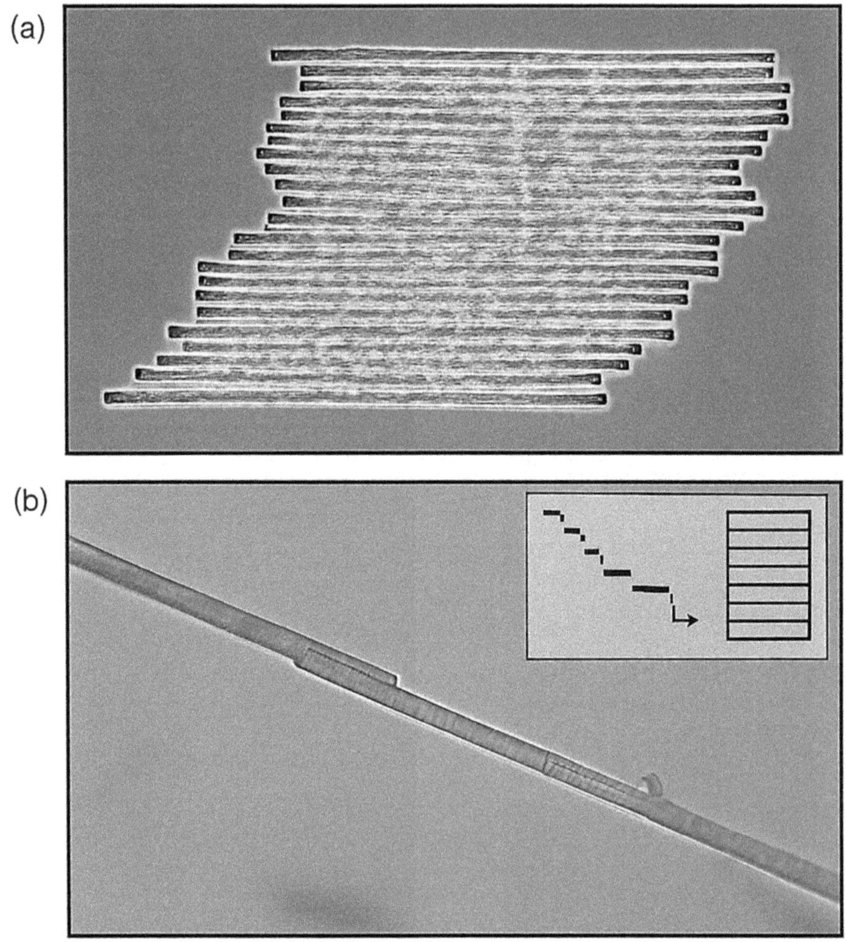

Figure 13.9 Light microscopy images of Bacillariapaxillifer. The single cells, which are about 100 μm long, slide against each other (see inset). The movement goes from a stack of cells (a) to an elongated band (b), back to the stack, and then to an elongation once more. Movies on B. paxillifer motion can be found on the internet.

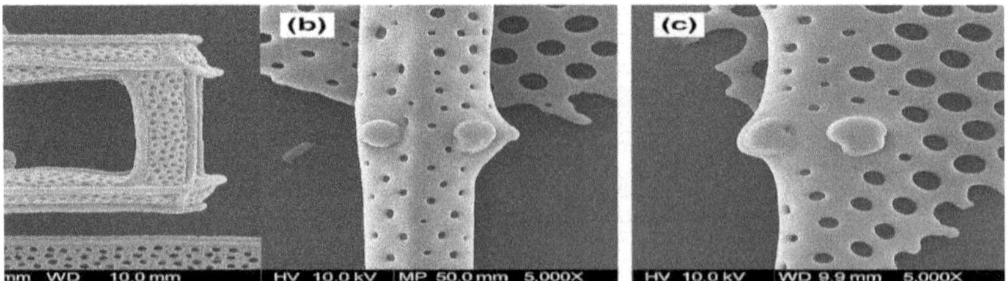

Figure 13.10 SEM images of an Eocene fossil (45 million years old) from a deposit at Mors, Denmark. (b) and (c) show the linking structures in more detail. Scale bars: 20 μm, 5 μm, and 5 μm, respectively.

13.3 CHALLENGES IN BIOTRIBOLOGY

13.3.1 Joint tribology

13.3.1.1 Challenges in joint tribology

Creating durable joint implants is a complex task that requires overcoming various challenges in the field of tribology. Improved comprehension of material interactions (metals, ceramics, polymers) is necessary to enhance compatibility and durability. Replicating the lubrication found in natural joints continues to be a complex challenge [59]. Understanding and addressing the issues of wear and friction in weight-bearing joints is of utmost importance to prevent failures and improve mobility. Ensuring long-term biocompatibility to prevent any negative reactions is also crucial. Efforts are directed toward enhancing implant design, materials, and lubrication mechanisms. Our goal is to increase the longevity, minimize damage, and enhance the overall well-being of patients. Ensuring the best possible match between materials and comprehending lubrication processes is crucial for improving the performance of joint replacements [16]. Minimizing wear and friction in weight-bearing joints is essential for preventing implant failures and improving mobility. In addition, the development of joint implants poses a significant challenge in terms of ensuring long-term biocompatibility. Our focus is on enhancing implant design, materials, and lubrication mechanisms to improve durability and ultimately benefit patients [60].

13.3.2 Skin tribology

13.3.2.1 Challenges in skin tribology

The study of skin tribology presents difficulties as it involves intricate interactions with various external factors such as clothing and environmental conditions. Appreciating the various ways in which skin interacts with friction calls for customized solutions. The complexity of skin properties, such as moisture, texture, and thickness, introduces variations. It is essential to tackle the difficulties associated with skin aging, diseases, and injuries, as they have a significant impact on skin tribology [61]. Understanding the effects of skincare products, textiles, and medical devices on skin friction and wear is crucial for ensuring user comfort and minimizing problems such as irritation and pressure sores. Developing precise measurement techniques for skin tribological properties under different conditions is a significant challenge that necessitates methodological advancements. Our goal is to tackle these challenges head-on, driving the development of groundbreaking products and treatments that enhance skin health and overall well-being. Through in-depth analysis of how the skin interacts with external factors and by addressing the unique characteristics of different skin types, we strive to deliver innovative solutions [62].

13.3.3 Oral tribology

13.3.3.1 Challenges in oral tribology

Understanding the complex interactions among dental materials, tissues, and saliva in the oral cavity can be quite challenging. The complex interplay between the different characteristics of dental materials and the ever-changing conditions in the mouth gives rise to a wide range of tribological phenomena. The impact of friction and wear on dental appliances, such as braces and implants, extends to both their functionality and the health of oral tissues [63]. Further investigation is needed to understand the impact of different diets, oral hygiene practices, and systemic health conditions on oral tribology. Continual obstacles involve the development of more efficient lubrication techniques, materials that are resistant to wear, and the minimization of friction in orthodontic treatments. Gaining knowledge about biofilms, salivary proteins, and enzymes is essential in order to effectively prevent dental problems such as caries and implant failures. Tackling these obstacles promotes the field of oral tribology, resulting in enhanced dental procedures, improved patient satisfaction, and superior oral health results [64].

13.3.4 Tribology of other biological systems

13.3.4.1 Challenges in tribology of other biological systems

Addressing the complexities of tribology in biological systems is of utmost importance for a wide range of applications in medicine, biomimetics, and engineering. Comprehending intricate interfaces, such as those found in the cardiovascular system, animal locomotion, and plant surfaces, requires an examination of surface topography, materials, and environmental factors [65]. Understanding the intricate aspects of lubrication in joints, tendons, and implants can greatly improve their durability and lifespan. Exploring the fascinating world of gecko feet and insect wings, which possess remarkable adhesive and self-cleaning properties, presents intriguing biomimetic design challenges. Investigating the durability of soil-dwelling creatures such as pangolins can provide valuable knowledge for the development of bio-inspired materials [66]. Tackling these obstacles in biological tribology propels fields, such as biomedicine, robotics, and materials science forward. It encourages the development of cutting-edge technologies and environmentally friendly solutions by emulating nature's processes. This leads to advancements in medical devices, engineering designs, and the creation of sustainable materials. Gaining a deep understanding of these intricate biological interactions and applying them across different fields is crucial for driving technological advancements and promoting sustainable development [67].

13.3.5 Cardiovascular physiology related to biotribology

13.3.5.1 Challenges in cardiovascular physiology related to biotribology

Exploring the intricate relationship between cardiovascular physiology and biotribology poses significant hurdles in comprehending heart-related ailments and enhancing medical devices. A thorough comprehension of friction, wear, and adhesion in dynamic blood flow is crucial for studying mechanical interactions within the cardiovascular system, specifically in artificial heart valves and intravascular stents [68]. Ensuring patient safety by developing non-adhesive surfaces on blood-contacting devices to prevent thrombus formation, even in the presence of anti-coagulation medication. Exploring the adhesive and frictional properties of contact lenses and their potential biological impacts, such as bacterial infections and discomfort, presents unique challenges in the field of ocular biotribology [69]. It is crucial to tackle these challenges in order to enhance cardiovascular implants and contact lens designs, resulting in better patient outcomes and an improved quality of life [70].

13.3.6 Biotribological model systems

13.3.6.1 Challenges in biotribological model systems

Developing successful biotribological model systems poses difficulties in reproducing intricate biological processes in laboratory environments, like the lubrication of synovial joints or interactions within the oral mucosa. Obtaining precise depictions necessitates creative methods and a thorough comprehension of biology [71]. Ensuring model reproducibility can be quite challenging, as even small experimental variations can result in a wide range of outcomes. Incorporating real-time physiological factors, such as temperature and humidity, can be challenging, but it is crucial for achieving accurate in vivo mimicry. It is essential to tackle these challenges in order to develop strong biotribological models, which will allow for the investigation of innovative solutions in the fields of medicine, materials science, and biotechnology [72]. Obtaining precise depictions of intricate biological interactions in laboratory settings, such as the lubrication of synovial joints or interactions within the oral mucosa, necessitates inventive methods and a profound comprehension of biological behaviors. Furthermore, it is essential to maintain consistency in model systems, even when faced with minor experimental differences, and to incorporate real-time physiological factors to accurately replicate in vivo situations. Addressing these obstacles is crucial in developing dependable biotribological models, pushing forward studies in medicine, materials science, and biotechnology [73].

REFERENCES

1. Gao, Leiming, Xianjiu Lu, Xiaogang Zhang, Qingen Meng, and Zhongmin Jin. "Lubrication modelling of artificial joint replacements: Current status and future challenges." *Lubricants* 10, no. 10 (2022): 238.
2. Ingale, Dhanashri, Priya Kulkarni, Ali Electricwala, Alpana Moghe, Sara Kamyab, Suresh Jagtap, Aare Martson, Sulev Koks, and Abhay Harsulkar. "Synovium-synovial fluid axis in osteoarthritis pathology: A key regulator of the cartilage degradation process." *Genes* 12, no. 7 (2021): 989.
3. Garg, T., G. Sharma, S. Shankar, and S. P. Singh. "Tribological performance of polymeric materials for biomedical applications." In *Assessment of Polymeric Materials for Biomedical Applications*, pp. 121–138. CRC Press, 2023.
4. Savin, Liliana, Tudor Pinteala, Dana Nicoleta Mihai, Dan Mihailescu, Smaranda Stefana Miu, Mihnea Theodor Sirbu, Bogdan Veliceasa, Dragos Cristian Popescu, Paul Dan Sirbu, and Norin Forna. "Updates on biomaterials used in total hip arthroplasty (THA)." *Polymers* 15, no. 15 (2023): 3278.
5. Hernandez, Ernesto, David Davidson, and Jacobo Rafael Reyes-Romero. "Characteristics of polymeric materials used in medicine." In *Materials for Biomedical Engineering*, pp. 479–506. Elsevier, 2019.
6. An, Haoming, Yubo Liu, Jiafeng Yi, Hongbin Xie, Chao Li, Xing Wang, and Wei Chai. "Research progress of cartilage lubrication and biomimetic cartilage lubrication materials." *Frontiers in Bioengineering and Biotechnology* 10 (2022): 1012653.
7. Mostakhdemin, Mohammad, Ashveen Nand, and Maziar Ramezani. "Articular and artificial cartilage, characteristics, properties and testing approaches—A review." *Polymers* 13, no. 12 (2021): 2000.
8. Chatterjee, Abhinava, Devendra K. Dubey, and Sujeet K. Sinha. "Effect of loading on the adhesion and frictional characteristics of top layer articular cartilage nanoscale contact: A molecular dynamics study." *Langmuir* 37, no. 1 (2020): 46–62.
9. Chatterjee, Abhinava, Devendra K. Dubey, and Sujeet K. Sinha. "Nanoscale friction and adhesion mechanisms in articular cartilage top layer hydrated interfaces: Insights from atomistic simulations." *Applied Surface Science* 550 (2021): 149216.
10. Wang, Mingkai, Yan Wu, Guangfeng Li, Qiushui Lin, Wencai Zhang, Han Liu, and Jiacan Su. "Articular cartilage repair biomaterials: Strategies and applications." *Materials Today Bio* 24 (2024): 100948.
11. Costa, Fábio Ramos, Márcia da Silva Santos, Rubens Andrade Martins, Cláudia Bruno Costa, Paulo César Hamdan, Marcos Britto Da Silva, Gabriel Ohana Marques Azzini et al. "The synergistic effects of hyaluronic acid and platelet-rich plasma for patellar chondropathy." *Biomedicines* 12, no. 1 (2023): 6.
12. Zelinka, Alena, Anke J. Roelofs, Rita A. Kandel, and Cosimo De Bari. "Cellular therapy and tissue engineering for cartilage repair." *Osteoarthritis and Cartilage* 12 (2022): 1547–1560.
13. Zhao, Xiaoduo, Weiyi Zhao, Yunlei Zhang, Xiaoqing Zhang, Zhengfeng Ma, Rui Wang, Qiangbing Wei, Shuanhong Ma, and Feng Zhou. "Recent progress of bioinspired cartilage hydrogel lubrication materials." *Biosurface and Biotribology* 8, no. 3 (2022): 225–243.

14. Soliman, Md Mohiuddin, Muhammad EH Chowdhury, Mohammad Tariqul Islam, Farayi Musharavati, Mohammad Nabil, Muhammad Hafizh, Amith Khandakar et al. "A review of biomaterials and associated performance metrics analysis in pre-clinical finite element model and in implementation stages for total hip implant system." *Polymers* 14, no. 20 (2022): 4308.
15. Little, Dianne, Peter C. Amadio, Hani A. Awad, Stephanie G. Cone, Nathaniel A. Dyment, Matthew B. Fisher, Alice H. Huang et al. "Preclinical tendon and ligament models: Beyond the 3Rs (replacement, reduction, and refinement) to 5W1H (why, who, what, where, when, how)." *Journal of Orthopaedic Research* 41, no. 10 (2023): 2133–2162.
16. Merola, Massimiliano, and Saverio Affatato. "Materials for hip prostheses: A review of wear and loading considerations." *Materials* 12, no. 3 (2019): 495.
17. Milošev, Ingrid, Vesna Levašič, Simon Kovač, Tarvo Sillat, Sannakaisa Virtanen, Veli-Matti Tiainen, and Rihard Trebše. "Metals for joint replacement." In *Joint Replacement Technology*, pp. 65–122. Woodhead Publishing, 2021.
18. Lee, Youngoh, Jonghwa Park, Ayoung Choe, Seungse Cho, Jinyoung Kim, and Hyunhyub Ko. "Mimicking human and biological skins for multifunctional skin electronics." *Advanced Functional Materials* 30, no. 20 (2020): 1904523.
19. Göçerler, Hakan, Carsten Gachot, Philipp G. Grützmacher, and Stefan J. Eder. "Skin as an interface: Understanding the synergy of dermatology, biomimetics and tribology." *Tribology and Materials* 2, no. 3 (2023): 128–153.
20. Ferrari, Michele, Francesca Cirisano, and M. Carmen Morán. "Mammalian cell behavior on hydrophobic substrates: Influence of surface properties." *Colloids and Interfaces* 3, no. 2 (2019): 48.
21. Zhang, Wengang, Jiaying Xiang, Ruijie Huang, and Hanlong Liu. "A review of bio-inspired geotechnics-perspectives from geomaterials, geo-components, and drilling & excavation strategies." *Biogeotechnics* 1, no. 3 (2023): 100025.
22. Li, Xuefei, Zhiwei Guo, Qiren Huang, and Chengqing Yuan. "Application of bionic tribology in water-lubricated bearing: A review." *Journal of Bionic Engineering* 19, no. 4 (2022): 902–934.
23. Al-Mahdi, Mohammed Riyadh Abdulameer. "Influence of pre-rolling process on the microstructure, mechanical and shape memory properties of Cu-Al-Ni shape memory alloys." PhD diss., 2022.
24. Zapoozhets, T., E. Tkachenko, and Svitlana Tryniak. "Dental physiology." (2019).
25. Wetselaar, Peter, Frank Lobbezoo, Ralph de Vries, Shamir B. Mehta, Niek J.M. Opdam, and Bas A.C. Loomans. "Developing diagnostic criteria for tooth wear, a preliminary beta version based on expert opinion, and a narrative literature review." *Journal of Oral Rehabilitation* 50, no. 10(2023): 1030–1042.
26. Beatriz Blancato, Ariane, Mariana Martins Bruno, Marina Faria Mendonça, and César Penazzo Lepri. "Intrinsic and extrinsic dental erosion: Literature review." *RSBO: Revista Sul-Brasileira de Odontologia* 20, no. 1 (2023).
27. Kruzic, Jamie J., Mark Hoffman, and Joseph A. Arsecularatne. "Fatigue and wear of human tooth enamel: A review." *Journal of the Mechanical Behavior of Biomedical Materials* 138 (2023): 105574.
28. Wetselaar, Peter, Daniele Manfredini, Jari Ahlberg, Anders Johansson, Ghizlane Aarab, Chryssa E. Papagianni, Marisol Reyes Sevilla, Michail Koutris, and Frank Lobbezoo. "Associations between tooth wear and dental sleep disorders: A narrative overview." *Journal of Oral Rehabilitation* 46, no. 8 (2019): 765–775.

29. Heintze, Siegward D., Franz-Xaver Reichl, and Reinhard Hickel. "Wear of dental materials: Clinical significance and laboratory wear simulation methods—A review." *Dental Materials Journal* 38, no. 3 (2019): 343–353.
30. Somya, Wilson. "Effect of silver and titanium nanocoating on frictional properties of stainless steel archwire: An in vitro study." PhD diss., Chettinad Dental College & Research Institute, Kanchipuram, 2020.
31. Selvaraj, Madhanraj, Kaja Mohaideen, Karthik Sennimalai, Greeshma Shantharam Gothankar, and Garima Arora. "Effect of oral environment on contemporary orthodontic materials and its clinical implications." *Journal of Orthodontic Science* 12 (2023).
32. Zeng, Qihang, Genlei Ma, Heng Xiao, Dan Yang, Jing Zheng, Liang Zheng, and Zhongrong Zhou. "Effect of saliva flow rate on the adsorption kinetics and lubrication of salivary pellicle on human tooth enamel surface." *Wear* 426 (2019): 180–185.
33. Zhai, Wenzheng, Lichun Bai, Runhua Zhou, Xueling Fan, Guozheng Kang, Yong Liu, and Kun Zhou. "Recent progress on wear-resistant materials: designs, properties, and applications." *Advanced Science* 8, no. 11 (2021): 2003739.
34. Williams, David F. "Assessing the triad of biocompatibility, medical device functionality and biological safety." *Medical Devices & Sensors* 4, no. 1 (2021): e10150.
35. McMonnies, Charles W. "Mechanisms for the symptoms of dryness in soft contact lens wearers." *Expert Review of Ophthalmology* 18, no. 5 (2023): 327–335.
36. Badini, Silvia, Stefano Regondi, and Raffaele Pugliese. "Unleashing the power of artificial intelligence in materials design." *Materials* 16, no. 17 (2023): 5927.
37. Ge-Zhang, Shangjie, Taoyang Cai, Hong Yang, Yuyang Ding, and Mingbo Song. "Biology and nature: Bionic superhydrophobic surface and principle." *Frontiers in Bioengineering and Biotechnology* 10 (2022): 1033514.
38. Evans, Jonathan T., Jonathan P. Evans, Robert W. Walker, Ashley W. Blom, Michael R. Whitehouse, and Adrian Sayers. "How long does a hip replacement last? A systematic review and meta-analysis of case series and national registry reports with more than 15 years of follow-up." *The Lancet* 393, no. 10172 (2019): 647–654.
39. Moghadasi, Kaveh, Mohammad Syahid Mohd Isa, Mohammad Ashraf Ariffin, Sufian Raja, Bo Wu, Mehrdad Yamani, Mohd Ridha Bin Muhamad et al. "A review on biomedical implant materials and the effect of friction stir based techniques on their mechanical and tribological properties." *Journal of Materials Research and Technology* 17 (2022): 1054–1121.
40. Arulkumar, M., R. Prashanna Rangan, M. Prem Ananth, V. Srividhyasakthi, and R. Aaditya. "Experimental verification on the influence of surface texturing on biomaterials and study of its tribological characteristics." *Materials Today: Proceedings* (2023).
41. Johns-Rahnejat, Patricia M., Ramin Rahmani, and Homer Rahnejat. "Current and future trends in tribological research." *Lubricants* 11, no. 9 (2023): 391.
42. Bennett, Elena M., Wolfgang Cramer, Alpina Begossi, Georgina Cundill, Sandra Díaz, Benis N. Egoh, Ilse R. Geijzendorffer et al. "Linking biodiversity, ecosystem services, and human well-being: Three challenges for designing research for sustainability." *Current Opinion in Environmental Sustainability* 14 (2015): 76–85.

43. Accioni, Francesca, Juan Vázquez, Manuel Merinero, Belén Begines, and Ana Alcudia. "Latest trends in surface modification for dental implantology: Innovative developments and analytical applications." *Pharmaceutics* 14, no. 2 (2022): 455.
44. Vasamsetty, Praveen, Tejaswini Pss, Divya Kukkala, Madhavi Singamshetty, and Shashivardhan Gajula. "3D printing in dentistry–Exploring the new horizons." *Materials Today: Proceedings* 26 (2020): 838–841.
45. Gebeshuber, Ilse C., Burhanuddin Yeop Majlis, and Herbert Stachelberger. "Tribology in biology: biomimetic studies across dimensions and across fields." *International Journal of Mechanical and Materials Engineering* 4, no. 3 (2009): 321–327.
46. Dowson, Duncan. "Bio-tribology." *Faraday Discussions* 156, no. 1 (2012): 9–30.
47. Piscitelli, Prisco, Giovanni Iolascon, Massimo Innocenti, Roberto Civinini, Alessandro Rubinacci, Maurizio Muratore, Michele D'Arienzo, Paolo Tranquilli Leali, Anna Maria Carossino, and Maria Luisa Brandi. "Painful prosthesis: approaching the patient with persistent pain following total hip and knee arthroplasty." *Clinical Cases in Mineral and Bone Metabolism* 10, no. 2 (2013): 97.
48. Kennedy, Simon, and Rhian M. Touyz. "Artery Structure–4 The Endothelium–5." *Textbook of Vascular Medicine* 1 (2019): 2.
49. Krüger-Genge, Anne, Anna Blocki, Ralf-Peter Franke, and Friedrich Jung. "Vascular endothelial cell biology: An update." *International Journal of Molecular Sciences* 20, no. 18 (2019): 4411.
50. Cornelissen, Anne, and Felix Jan Vogt. "The effects of stenting on coronary endothelium from a molecular biological view: Time for improvement?." *Journal of Cellular and Molecular Medicine* 23, no. 1 (2019): 39–46.
51. Hohmann, Tim, and Faramarz Dehghani. "The cytoskeleton—A complex interacting meshwork." *Cells* 8, no. 4 (2019): 362.
52. Gordon, Emma, Lilian Schimmel, and Maike Frye. "The importance of mechanical forces for in vitro endothelial cell biology." *Frontiers in Physiology* (2020): 684.
53. Wagner, Rasmus MF, Raman Maiti, Matt J. Carré, Cécile M. Perrault, Paul C. Evans, and Roger Lewis. "Bio-tribology of vascular devices: A review of tissue/device friction research." *Biotribology* 25 (2021): 100169.
54. Badv, Maryam, Fereshteh Bayat, Jeffrey I. Weitz, and Tohid F. Didar. "Single and multi-functional coating strategies for enhancing the biocompatibility and tissue integration of blood-contacting medical implants." *Biomaterials* 258 (2020): 120291.
55. Gan, Jinping, Brad Bolon, Terry Van Vleet, and Charles Wood. "Alternative models in biomedical research: In silico, in vitro, ex vivo, and nontraditional in vivo Approaches. In *Haschek and Rousseaux's Handbook of Toxicologic Pathology*, pp. 925–966. Academic Press, 2022.
56. Sardo, Angela, Ida Orefice, Sergio Balzano, Lucia Barra, and Giovanna Romano. "Mini-review: Potential of diatom-derived silica for biomedical applications." *Applied Sciences* 11, no. 10 (2021): 4533.
57. Sun, Xiaojie, Na Li, Chang Su, Yuzhi Mu, Xin Cong, Zheng Cao, Xiaoye Wang, Xiaoyan Yang, Xiguang Chen, and Chao Feng. "Diatom-inspired bionic hydrophilic polysaccharide adhesive for rapid sealing hemostasis." *ACS Nano* 17, no. 19 (2023): 19121–19135.

58. Tahmooresnejad, Leila, and Catherine Beaudry. "Collaboration or funding: Lessons from a study of nanotechnology patenting in Canada and the United States." *The Journal of Technology Transfer* 44 (2019): 741–777.
59. Chen, Zhangwei, Ziyong Li, Junjie Li, Chengbo Liu, Changshi Lao, Yuelong Fu, Changyong Liu, Yang Li, Pei Wang, and Yi He. "3D printing of ceramics: A review." *Journal of the European Ceramic Society* 39, no. 4 (2019): 661–687.
60. Bairagi, Darothi, and Sumantra Mandal. "A comprehensive review on biocompatible Mg-based alloys as temporary orthopaedic implants: Current status, challenges, and future prospects." *Journal of Magnesium and Alloys* 10, no. 3 (2022): 627–669.
61. D'Souza, Brian, Ashish K. Kasar, Jaycob Jones, Andre Skeete, Lane Rader, Pankaj Kumar, and Pradeep L. Menezes. "A brief review on factors affecting the tribological interaction between human skin and different textile materials." *Materials* 15, no. 6 (2022): 2184.
62. Schwingen, Julius, Mustafa Kaplan, and Florian C. Kurschus. "Current concepts in inflammatory skin diseases evolved by transcriptome analysis: In-depth analysis of atopic dermatitis and psoriasis." *International Journal of Molecular Sciences* 21, no. 3 (2020): 699.
63. Smart, Pravin Alexander. "Mucin layer behaviour and its role in dental tribology." PhD diss., University of Leeds, 2022.
64. Li, Gefei, Qiang Yu, Mengqi Li, Dingkun Zhang, Ji Yu, Xiaohan Yu, Chenxi Xia, Junzhi Lin, Li Han, and Haozhou Huang. "*Phyllanthus emblica* fruits: A polyphenol-rich fruit with potential benefits for oral management." *Food & Function* 14, no. 17 (2023): 7738–7759.
65. Pradal, Clementine, Gleb E. Yakubov, Martin A.K. Williams, Michael A. McGuckin, and Jason R. Stokes. "Lubrication by biomacromolecules: Mechanisms and biomimetic strategies." *Bioinspiration & Biomimetics* 14, no. 5 (2019): 051001.
66. Primrose, Sandy B. *Biomimetics: Nature-Inspired Design and Innovation.* John Wiley & Sons, 2020.
67. Palomares, Iván, Eugenio Martínez-Cámara, Rosana Montes, Pablo García-Moral, Manuel Chiachio, Juan Chiachio, Sergio Alonso et al. "A panoramic view and swot analysis of artificial intelligence for achieving the sustainable development goals by 2030: Progress and prospects." *Applied Intelligence* 51 (2021): 6497–6527.
68. Wang, Lizhen, Xili Ding, Wentao Feng, Yuanming Gao, Shudong Zhao, and Yubo Fan. "Biomechanical study on implantable and interventional medical devices." *Acta Mechanica Sinica* 37, no. 6 (2021): 875–894.
69. Nečas, David, Vlastimil Kulíšek, Petr Štěpán, František Ondreáš, Pavel Čípek, Gloria Huerta-Angeles, and Martin Vrbka. "Friction and lubrication of eye/lens/lid interface: The effect of lubricant and contact lens material." *Tribology Letters* 71, no. 4 (2023): 118.
70. Hübner, Constanze, Mariya Lorke, Annika Buchholz, Stefanie Frech, Laura Harzheim, Sabine Schulz, Saskia Jünger, and Christiane Woopen. "Health literacy in the context of implant care—Perspectives of (prospective) implant wearers on individual and organisational factors." *International Journal of Environmental Research and Public Health* 19, no. 12 (2022): 6975.

71. Hirwani, Jaswant Kumar, and Sujeet K. Sinha. "Bio-tribological studies of Structalit/UHMWPE composites as an alternative to UHMWPE for hip joint application." *Wear* 518 (2023): 204630.
72. Rao, T. V. V. L. N., S. B. Kasolang, X. Guoxin, J. K. Katiyar, and A. M. A. Rani, eds. *Biotribology: Emerging Technologies and Applications*. CRC Press, 2021.
73. Berry, Matthew E., Samantha M. McCabe, Sian Sloan-Dennison, Stacey Laing, Neil C. Shand, Duncan Graham, and Karen Faulds. "Tomographic imaging and localization of nanoparticles in tissue using surface-enhanced spatially offset Raman spectroscopy." *ACS Applied Materials & Interfaces* 14, no. 28 (2022): 31613–31624.

Index

Pages in *italics* refer to figures and pages in **bold** refer to tables.

For Product Safety Concerns and Information please contact our EU representative GPSR@taylorandfrancis.com
Taylor & Francis Verlag GmbH, Kaufingerstraße 24, 80331 München, Germany

www.ingramcontent.com/pod-product-compliance
Ingram Content Group UK Ltd.
Pitfield, Milton Keynes, MK11 3LW, UK
UKHW020200200926
475702UK00004B/544